CONGRESSIONAL POLICIES, PRACTICES AND PROCEDURES

COAST GUARD PROGRAMS

IMPLEMENTATION, AUTHORIZATION AND FISCAL REQUIREMENTS

Congressional Policies, Practices and Procedures

Additional books and e-books in this series can be found on Nova's website under the Series tab.

CONGRESSIONAL POLICIES, PRACTICES AND PROCEDURES

COAST GUARD PROGRAMS

IMPLEMENTATION, AUTHORIZATION AND FISCAL REQUIREMENTS

DAVID J. SIMMONS
EDITOR

snova
New York

NOTICE TO THE READER

The Publisher has taken reasonable care in the preparation of this book, but makes no expressed or implied warranty of any kind and assumes no responsibility for any errors or omissions. No liability is assumed for incidental or consequential damages in connection with or arising out of information contained in this book. The Publisher shall not be liable for any special, consequential, or exemplary damages resulting, in whole or in part, from the readers' use of, or reliance upon, this material. Any parts of this book based on government reports are so indicated and copyright is claimed for those parts to the extent applicable to compilations of such works.

Independent verification should be sought for any data, advice or recommendations contained in this book. In addition, no responsibility is assumed by the Publisher for any injury and/or damage to persons or property arising from any methods, products, instructions, ideas or otherwise contained in this publication.

This publication is designed to provide accurate and authoritative information with regard to the subject matter covered herein. It is sold with the clear understanding that the Publisher is not engaged in rendering legal or any other professional services. If legal or any other expert assistance is required, the services of a competent person should be sought. FROM A DECLARATION OF PARTICIPANTS JOINTLY ADOPTED BY A COMMITTEE OF THE AMERICAN BAR ASSOCIATION AND A COMMITTEE OF PUBLISHERS.

Additional color graphics may be available in the e-book version of this book.

Library of Congress Cataloging-in-Publication Data

ISBN: 978-1-53615-426-9

Published by Nova Science Publishers, Inc. † New York

CONTENTS

PREFACE

Chapter 1 examines the implementation of certain Coast Guard programs, including those involving performance monitoring, the Service's Capital Investment Plan, and commercial fishing vessel safety. Chapter 2 discusses Coast Guard and maritime transportation programs. Chapter 3 reviews the fiscal year 2019 budget request for the Coast Guard and maritime transportation programs

Chapter 1 - This is an edited, reformatted and augmented version of Hearing Before the Subcommittee on Coast guard and Maritime Transportation of the Committee on Transportation and Infrastructure, House Of Representatives, One Hundred Fifteenth Congress, Second Session, dated March 7, 2018. This hearing builds upon the Subcommittee's previous hearings focused on the Service's mission performance and long-term planning for the acquisition of major systems. Those hearings included extensive reviews of Coast Guard acquisition programs, as well as reviews of the Service's workforce composition and shoreside infrastructure needs. Additionally, prior hearings examined the Coast Guard's operational planning and its development of objective goals and metrics to track and measure performance. This hearing furthers the Subcommittee's ongoing oversight of these important Coast Guard activities. In addition, the hearing will allow the Subcommittee to further long-standing efforts to improve maritime safety in the fishing industry by receiving input on the implications of policy changes made in 2010 and 2012 affecting the construction of commercial fishing vessels.

Chapter 2 - This is an edited, reformatted and augmented version of Hearing before the Subcommittee on Coast Guard and Maritime Transportation, of the Committee on Transportation and Infrastructure, House of Representatives, One Hundred Fifteenth Congress, First Session, dated April 4, 2017. Each Congress the Subcommittee on Coast Guard and Maritime Transpot1ation moves an authorization bill principally to authorize appropriations for the Coast Guard and also the FMC. The last reauthorization bill (the *Coast Guard Authorization Act of 2016,* P.L. 114-120), provided a two-year authorization of appropriations for fiscal years (FY) 2016 and 2017, with $9.1 billion authorized for each year. This Congress, we will need to move another authorization bill. Such bill may include: provisions suggested by the Coast Guard to improve its operations and programs; studies or reports to be conducted by the Government Accountability Office to review laws and regulations implemented by the Service; legislative changes to maritime laws; land conveyances; and miscellaneous maritime policy items. Historically, the Subcommittee works with the House Armed Services Committee to authorize appropriations for MARAD in the National Defense Authorization bill which is passed each session of Congress. The Subcommittee retains jurisdiction over non-national security aspects of the merchant marine and when appropriate addresses relevant issues in authorization bills.

Chapter 3 - This is an edited, reformatted and augmented version of Hearing before the Subcommittee on Coast Guard and Maritime Transportation of the Committee on Transportation and Infrastructure, House of Representatives, One Hundred Fifteenth Congress, Second Session, dated March 14, 2018. The Coast Guard was established on January 28, 1915, through the consolidation of the Revenue Cutter Service (established in 1790) and the Lifesaving Service (established in 1848). The Coast Guard later assumed the duties of three other agencies: the Lighthouse Service (established in 1789), the Steamboat Inspection Service (established in 1838), and the Bureau of Navigation (established in 1884). Under Section 2 of Title 14, United States Code, the Coast Guard has primary responsibility to enforce or assist in the enforcement of all applicable federal laws on, under, and over the high seas and waters subject to the jurisdiction of the United States; to ensure safety of life and property at sea; to carry out domestic and international icebreaking activities; and, as one of the five armed forces of the United States, to maintain defense readiness to operate as a specialized service in the Navy upon the declaration of war or when the President directs. The Coast Guard is directed by a Commandant, who is appointed by the President with the advice and consent of the Senate to a four-year term. On May 30, 2014, President Obama appointed Admiral Paul F. Zukunft as Commandant of the Coast Guard. His term expires in May 2018. On March 8, 2018, President Trump announced his intention to nominate Vice Admiral Karl Schultz as the 26th Commandant of the Coast Guard.

In: Coast Guard Programs
Editor: David J. Simmons

ISBN: 978-1-53615-426-9

Chapter 1

IMPLEMENTATION OF COAST GUARD PROGRAMS*

Subcommittee on Coast Guard and Maritime Transportation

SUMMARY OF SUBJECT MATTER

TO: Members, Subcommittee on Coast Guard and Maritime Transportation
FROM: Staff, Subcommittee on Coast Guard and Maritime Transportation
RE: Hearing on "Implementation of Coast Guard Programs"

PURPOSE

The Subcommittee on Coast Guard and Maritime Transportation will hold a hearing on Wednesday, March 7, 2018, at 2:00 p.m., in 2167 Rayburn House Office Building to examine the implementation of certain Coast Guard Programs. The Subcommittee will also examine a new report investigating the cost implications of building fishing vessels to classification standards. Witnesses from the United States Coast Guard (Coast Guard or Service) and the Government Accountability Office (GAO) will testify.

* This is an edited, reformatted and augmented version of Hearing Before the Subcommittee on Coast guard and Maritime Transportation of the Committee on Transportation and Infrastructure, House Of Representatives, One Hundred Fifteenth Congress, Second Session, dated March 7, 2018.

BACKGROUND

This hearing builds upon the Subcommittee's previous hearings focused on the Service's mission performance and long-term planning for the acquisition of major systems. Those hearings included extensive reviews of Coast Guard acquisition programs, as well as reviews of the Service's workforce composition and shoreside infrastructure needs. Additionally, prior hearings examined the Coast Guard's operational planning and its development of objective goals and metrics to track and measure performance. This hearing furthers the Subcommittee's ongoing oversight of these important Coast Guard activities. In addition, the hearing will allow the Subcommittee to further long-standing efforts to improve maritime safety in the fishing industry by receiving input on the implications of policy changes made in 2010 and 2012 affecting the construction of commercial fishing vessels.

PERFORMANCE INFORMATION TRANSPARENCY AND MONITORING

The Coast Guard has 11 statutory missions (6 U.S.C. § 648): marine safety; search and rescue; aids to navigation; living marine resources (fisheries law enforcement); marine environmental protection; ice operations; ports, waterways and coastal security; drug interdiction; migrant interdiction; defense readiness, and other law enforcement. As required by the Coast Guard Authorization Act of 2015 (P,L. 114-120), the GAO recently completed a report entitled Actions Needed to Enhance Performance Information Transparency and Monitoring (GAO-18-13), addressing whether the Coast Guard's annual performance goals and reported performance information accurately reflects the extent to which the Service is accomplishing its mission responsibilities. GAO found the following:

- Goals representing five of the 11 missions do not fully address all related mission activities, and developing new or adjusting existing goals could better convey the Coast Guard's performance.
- The Coast Guard does not sufficiently report its performance goals and actual performance in publicly available documents, limiting congressional and public awareness of the Service's ability to meet its missions.

The Coast Guard is a complex organization and accurately capturing performance across all of its missions is a complicated endeavor. Nevertheless, it is critical that the

Service establish performance measures that address all mission activities and that actual performance be reported accurately and routinely to Congress and the public.

Five-Year Capital Investment Plan

Coast Guard mission requirements often require capital investment in assets and equipment (e.g., vessels, aircraft, shore infrastructure, information technologies, etc.), which should be reflected in its Capital Investment Plan (CIP). Section 2902 of title 14, United States Code, requires the Commandant of the Coast Guard to submit a CIP to the Committee on Transportation and Infrastructure each year in conjunction with the Administration's respective budget request. The CIP identifies projected funding levels over the next five fiscal years for each major acquisition, as well as estimated timelines and total costs to complete each such acquisition (note: a "major acquisition program" means an ongoing acquisition with a life-cycle cost estimate greater than or equal to $300 million). The purpose of the CIP is to ensure Congress has adequate information to conduct proper oversight of the Service's capital budget, acquisition plans, mission needs, and readiness to conduct operations in future years.

The Commandant has testified before the Subcommittee that the Service requires $2 billion per year in Acquisition, Construction and Improvement funds to meet the operational needs of the Service. The CIP does not provide for that level of investment. The Commandant has testified that the Service needs new cutters and aircraft, yet the CIP does not include funding to deliver new assets in a timely manner to replace aged legacy assets. Likewise, the Service consistently speaks of a major funding shortfall to build and repair shore infrastructure (the current shore infrastructure backlog is estimated at over $1.6 billion), yet the Coast Guard regularly requests only nominal funding for shore infrastructure projects (e.g., only $10 million requested in fiscal year (FY) 2018). Because the CIP does not accurately reflect current mission requirements, Congress is left to set priorities for the Coast Guard without appropriate administrative guidance.

The GAO has criticized Coast Guard CIPs for failing to accurately reflect cost and schedule impacts from funding shortfalls. The 2014 GAO report entitled *Better Information on Performance and Funding Needed to Address Shortfalls* (GA0-14-450), recommended that the Coast Guard be required to regularly update the estimated timelines and total costs to complete each acquisition based upon actual appropriations provided by Congress, as opposed to projected funding levels. The Coast Guard continues to under-deliver in these areas.

The Coast Guard submitted the FY 2018-2022 CIP on October 24, 2017, over five months after it should have been submitted with the President's budget. Even though Congress annually has withheld $85 million of operational funding until the CIP is received, the Coast Guard has failed to meet the deadline each year since the requirement was first implemented in FY 2013. In addition to being late, the C1P was less than helpful, as most of the Life Cycle Cost Estimates (LCCEs) that form the basis for the acquisition project cost estimates and schedules are outdated. The information provided does not align with reality. Without updates to the LCCEs, Congress is unable to analyze the over $7.6 billion in funding the Coast Guard plans to request in FY 2018 to FY 2022.

The following is a partial list of CIP deficiencies regarding the Coast Guard's major acquisition programs:

- National Security Cutter (NSC): The estimates, projections, and schedules are based on an LCCE completed in September 2014, despite the Coast Guard being appropriated funding for construction of a ninth NSC in December 2015 and awarding the construction contract for it in December 2016, and then receiving appropriations for long lead time materials for a tenth NSC in May 2017. The CIP, however, does not account for the impacts of adding additional vessels to the NSC fleet or of the $735 million appropriated for that purpose.
- Offshore Patrol Cutter (OPC): The estimates, projections, and schedules provided in the CIP are based on an LCCE from March 2012, even though the Coast Guard awarded the OPC Phase I construction contract in 2016. The CIP does not take into account the actual known production costs and schedules.
- Fast Response Cutter (FRC): The estimates, projections, and schedules are based on an LCCE from February 2015, despite awarding the FRC Phase II construction contract in February 2016. The CIP does incorporate the most recent cost and schedule information.
- HC-130J Aircraft: The estimates, projections, and schedules are based on an LCCE from November 2011, despite major changes to the program in the past six years. The CIP does not acknowledge the transfer of Coast Guard HC-130H aircraft to the Forest Service and the Coast Guard's receipt of HC-27J aircraft from the Air Force. Similarly, the CIP also fails to account for appropriations provided for several HC-130J aircraft which were not requested in the President's budget or accounted for in initial planning documents.

Between FY 2013 and FY 2017, Congress appropriated over $2 billion more for acquisitions than the Coast Guard requested to meet emerging needs and appropriately position the Service to meet all mission demands.

The Coast Guard's failure to ensure CIP investment levels accurately reflect mission requirements and to adjust acquisition documentation to reflect reality leaves Congress without reliable input and data on which to base future appropriations and perform proper oversight.

Commercial Fishing Vessel Classification

Commercial fishing vessels are uninspected vessels under U.S. law, meaning the Coast Guard generally does not have the authority to inspect the vessels during construction or maintenance. Since 1988, pursuant to the *Commercial Fishing Industry Vessel Safety Act of 1988* (P.L. 100-424), commercial fish processing vessels have been required to be built and maintained to classification society standards. Such standards include the use of a naval architect and a classification society surveyor; the use of classification society approved materials by the shipyard; and the classification society assessing vessel stability, assigning a loading mark, and issuing loading directions to the owner. A vessel built to these standards receives a certificate indicating compliance with classification standards.

The *Coast Guard Authorization Acts of 2010* (P.L. 111-281) and *2012* (P.L. 112-213) expanded the class requirements to catcher vessels and fish tender vessels. These Acts also directed the Coast Guard to develop an Alternate Safety Compliance Program (ASCP) for commercial fishing vessels between 50-79 feet, built before July 1, 2013, and 25 years or older. To date, the Coast Guard has not issued guidance or regulations for either the class or alternate safety compliance requirements. Instead, in 2016, the Coast Guard developed an Enhanced Oversight Program, using existing policy and authorities, focused on older non-classed fishing vessels that may have an increased risk of vessel and crew loss.

The *Coast Guard Authorization Act of 2015* established an additional safety compliance program for new-build vessels (at least 50 feet overall in length, and not more than 79 feet overall in length). Program requirements stipulate that:

(1) The vessel is designed by an individual licensed by a state as a naval architect or marine engineer, and the design incorporates standards equivalent to those prescribed by a classification society to which the Secretary has delegated authority under section 3316 or another qualified organization approved by the Secretary for purposes of this paragraph.

(2) Construction of the vessel is overseen and certified as being in accordance with its design by a marine surveyor of an organization accepted by the Secretary.

(3) The vessel—
   a) completes a stability test performed by a qualified individual;
   b) has written stability and loading instructions from a qualified individual that are provided to the owner or operator; and
   c) has an assigned loading mark.

(4) The vessel is not substantially altered without the review and approval of an individual licensed by a state as a naval architect or marine engineer before the beginning of such substantial alteration.

(5) The vessel undergoes a condition survey at least twice in five years, not to exceed three years between surveys, to the satisfaction of a marine surveyor of an organization accepted by the Secretary.

(6) The vessel undergoes an out-of-water survey at least once every five years to the satisfaction of a certified marine surveyor of an organization accepted by the Secretary.

Despite that requirement, the Coast Guard has not issued any guidance or regulations to implement this new ASCP. While the use of ASCPs provides greater flexibility and potential cost savings to owners of smaller commercial fishing vessels, the lack of regulatory action or issuance of guidance to implement these programs contributes to confusion among the owners of commercial fishing vessels and the shipbuilding industry. The Coast Guard is working through the regulatory process to update its existing regulations to address the classification requirements of the 2010, 2012, and 2015 Acts. The commercial fishing industry has raised concerns about the cost of building and maintaining fishing vessels to class. The Coast Guard hopes to complete this more formal policy by December 31, 2018.

The *Coast Guard Authorization of 2015* required GAO to review commercial fishing vessel safety, including the costs and benefits of classing commercial fishing vessels. The report entitled *Commercial Fishing Vessels: More Information Needed to Improve Classification Implementation (GAO-18- 16)*, notes that there was only a very small data set of six fishing vessels (four were fish processors already required to be built to class since 1988) that have been built under the classification system since 2013. The report recommends that the Coast Guard work with the National Institute for Occupational Health and Safety, the National Marine Fisheries Service, and other federal agencies to collect reliable data on the number of active U.S. commercial fishing vessel and the fisheries in which they operate, as well as key vessel characteristics, including, but are not limited to, vessel age and length. It recommends that the data collected be used to assess commercial fishing vessel accidents, injuries, and fatalities. GAO also recommends that the Coast Guard develop regulations or guidance to address questions concerning the classification and the ASCP approaches to the design, construction, and maintenance of fishing vessels.

WITNESS LIST

Rear Admiral Linda Fagan

Deputy Commandant for Operations, Policy, and Capabilities United States Coast Guard

Mr. Nathan Anderson Acting Director Homeland Security and Justice Government Accountability Office

Wednesday, March 7, 2018
House of Representatives,
Subcommittee on Coast Guard
And Maritime Transportation,
Committee on Transportation and Infrastructure,
Washington, DC.

The subcommittee met, pursuant to call, at 2:04 p.m., in room 2167, Rayburn House Office Building, Hon. Duncan Hunter (Chairman of the subcommittee) presiding.

Mr. HUNTER. The subcommittee will come to order.

Good afternoon. Without objection, the Chair is authorized to declare a recess at any time.

The subcommittee is convening today to examine the implementation of certain Coast Guard programs, including those involving performance monitoring, the Service's Capital Investment Plan, and commercial fishing vessel safety.

It is important that Congress understands the Coast Guard's ability to meet its missions. The Coast Guard conducts search and rescue, drug interdiction, and defense readiness activities on a daily basis, yet it is unclear how the Service tracks these activities. Members of the committee rely on the Service to report its performance to Congress and to the public.

To ensure a full understanding of the Service's abilities, the Coast Guard Authorization Act of 2016 required the Government Accountability Office to look at whether the Coast Guard's annual performance goals and reported performance information accurately reflects the extent to which the Service is accomplishing its mission responsibilities. GAO found that the Coast Guard goals do not address all mission activities and that the Service does not sufficiently report its performance goals and actual performance.

In addition to recognizing the degree to which the Coast Guard is able to accomplish its missions today, it is also critical that the committee understands how the Service plans to do so in the future. The Capital Investment Plan is supposed to reflect the capital investments necessary to meet mission requirements in the future. Unfortunately, the Coast Guard submits the CIP late every year, precluding its use to properly inform appropriation and authorization legislation.

When a CIP is finally delivered, its profile often does not align with the needs of the Service as reported by the Commandant and other senior leaders. The Coast Guard says it needs new cutters and aircraft, yet the CIP contains outdated information that does not include funding to match those stated needs.

As a glaring example, despite an over $1.5 billion funding shortfall to build and maintain shore infrastructure, the Coast Guard requests only nominal funding, $10 million—so I will say that again, $1.5 billion, as stated by the Commandant and other senior leaders—and the Coast Guard requested $10 million in 2018 and $30 million for fiscal year 2019 for shore infrastructure projects.

Rather than providing the information necessary to understand and support Service priorities, the Coast Guard relies on Congress to set those priorities by appropriating funding above the requested levels. Unfortunately, the Coast Guard has failed to adjust acquisition documents to reflect the funding for additional assets, lessening the reliability and value of the information provided in the CIP.

The Coast Guard's failure to ensure CIP investment levels accurately reflect its mission requirements and prior appropriations leaves Congress without reliable information on which to base future appropriations and perform proper oversight.

One Coast Guard mission area on which Congress has provided significant direction is commercial fishing and vessel safety. Congressional requirements regarding the construction and maintenance of commercial fishing vessels were enacted in 1988 and expanded in 2010, 2012, and 2016. In addition, Congress directed the Coast Guard to implement regulations regarding vessel classification requirements and establish alternate safety compliance programs to allow smaller and older vessels to comply with requirements more suited to their unique characteristics.

Unfortunately, the Coast Guard has failed to act on those directives. Instead, in 2016 the Coast Guard developed a, quote, "Enhanced Oversight Program" using existing policies and authorities focusing on older, nonclassified, nonclass fishing vessels that may have had an increased risk of vessel and crew loss.

At the same time, the committee heard concerns from the commercial fishing industry regarding the lack of communication from the Coast Guard on the development of required alternate safety compliance programs. The Enhanced Oversight Program did not alleviate those concerns due to questions about the program being permanent or temporary, and if temporary, whether the Coast Guard would work with industry to develop the alternate safety compliance programs required by law.

GAO reviewed the Coast Guard's implementation of commercial fishing vessel classification requirements and its impact on new vessel construction and will provide testimony on their findings today. I look forward to hearing from the Coast Guard on where the Service stands on implementing all of these requirements and more that I have mentioned.

I thank our witnesses for being here today, and I look forward to hearing their thoughts on the issues.

Mr. Garamendi.

Mr. GARAMENDI. Thank you, Mr. Chairman. You have covered all of the issues that are before the committee and the concerns that the committee has. I would, therefore, like to for the record put my statement in. It deals with not only the issues you raised, but some of the other things that are before us, which are probably not going to be taken up today because of the time element that we have.

Without a doubt, for us to do our job of authorization as well as oversight we need to have a solid foundation. Mr. Chairman, you went through that not only with the CIP, but also with the fishing vessel issue.

So let's get on with it. My statement will be in the record for anybody that cares to search through the history.

Mr. HUNTER. Without objection.

Mr. GARAMENDI. Thank you. I yield back.

[Congressman Garamendi's prepared statement is on pages 25–27.]

Mr. HUNTER. I thank the ranking member.

Today we will hear testimony from Rear Admiral Linda Fagan, Deputy Commandant for Operations Policy and Capability, United States Coast Guard, and Mr. Nathan Anderson, Acting Director of Homeland Security and Justice, Government Accountability Office.

Rear Admiral Fagan, you are recognized to give your statement.

TESTIMONY OF REAR ADMIRAL LINDA L. FAGAN, DEPUTY COMMANDANT FOR OPERATIONS POLICY AND CAPABILITY, U.S. COAST GUARD; AND NATHAN ANDERSON, ACTING DIRECTOR, HOMELAND SECURITY AND JUSTICE, U.S. GOVERNMENT ACCOUNTABILITY OFFICE

Admiral FAGAN. Thank you, Mr. Chairman.

Good afternoon, Mr. Chairman, Ranking Member Garamendi, and distinguished members of the subcommittee. I appreciate the opportunity to testify today and ask that my written testimony please be accepted into the record.

Mr. HUNTER. Without objection.

Admiral FAGAN. Thank you for the enduring support and trust that Congress and this subcommittee have provided to the United States Coast Guard. The recent supplemental appropriation for hurricane response allows the Service to rebuild damaged infrastructure and remain resilient into the future. The Coast Guard is an integral part of many of these communities that suffered damage, and we were also impacted, as were the communities, and are committed to continuing to execute our missions in these regions.

The Coast Guard operates in an increasingly complex world and strives to make the best use of the limited resources we have at our disposal. Through a deliberate approach known as the Standard Operational Planning Process, we continuously evaluate maritime threats and opportunities and develop plans to achieve mission success.

Using national, departmental, and service strategies as guideposts, we leverage the intelligence community and this planning process to employ a risk-based approach to prioritize assets where they are needed most. Tactical commanders benefit from this process and have the flexibility to allocate resources on scene, ultimately allowing us to address maritime threats with the greatest precision and effect.

The planning process is guided by an annual strategic review to assess performance with robust metrics, identify operational gaps, and delineate steps needed to close them. We are appreciative of the GAO's recent review of the performance assessment system, and we are working to incorporate the recommendations of continual improvement in this regard. And the agility of this system was applied during the hurricane responses this past summer when we mobilized and deployed nearly 3,000 members and helped rescue over 11,000 people in need.

The performance measures we will discuss today are a result of a responsive, transparent, repeatable planning process. We shouldn't lose sight of the danger these measures reflect to our men and women who conduct frontline operations, and the decisions we make based on these metrics drive operational change in the field that affect our Coast Guard members. These are not just members, there are both public and Coast Guard lives on the line in these mission sets.

This risk is always present with me as I work on policy and resourcing here in Washington, and I would like to share with you a story of how the system has impacted just myself personally.

My daughter is a junior officer serving on one of our legacy cutters, a 210-foot cutter. It is 53 years old, out of Port Angeles, Washington. And this past October the *Active* was deployed into the Eastern Pacific in support of our Western Hemisphere strategy.

An aircraft detected two pangas that appeared to be engaged in illicit activity. Three o'clock in the morning the Coast Guard cutter *Active* launched their small boat to intercept and interdict the two pangas. One panga remained stationary, but the other fled with contraband on board, and the cutter small boat gave chase, high- speed chase, as the smugglers were discharging contraband overboard.

In the darkness and at high speed the cutter small boat and the panga collided. Following the collision, the panga swiftly surrendered and the *Active*'s crew successfully apprehended the smugglers and more than 800 kilograms of cocaine.

Thankfully, no one was hurt in the incident, but for me and for all of us the case is symbolic of the operations that the Service undertakes 24/7/365, and highlights the risk our men and women face in efforts to secure our borders, stop criminals, and save lives.

The *Active* was directed to the Eastern Pacific as a result of this operational planning system and our Western Hemisphere strategy provided the framework for operations, and intelligence helped showcase the threat in this particular asset. And it was an operational commander that directed my daughter's ship to face this dangerous mission set.

The Coast Guard's strategic planning process works. History has proven that a responsive, capable, and agile Coast Guard, using a deliberate planning process, is an indispensable instrument of national security.

To protect American security and economic interests the Coast Guard continues to improve its strategic planning system. We look forward to continue to invest smartly in capabilities, maintain our technological edge, recruit, develop, and retain a highly capable workforce in order to succeed in this increasingly complex environment.

Thank you again for the opportunity to testify before you today, and I look forward to your questions.

Mr. HUNTER. Thank you, Admiral.

Mr. Anderson, I understand this is your first time.

Mr. ANDERSON. Yes, sir.

Mr. HUNTER. Good. Go ahead.

Mr. ANDERSON. Chairman Hunter, Ranking Member Garamendi, and members of the subcommittee, good afternoon.

My testimony today is primarily based on work we have conducted over the past few years. I will address Coast Guard actions needed to improve the quality of data used for program management and improve the transparency of information used for reporting on its mission performance and capital planning.

With regard to data quality, we have issued three reports over the past 2 years which underscore the need to collect and use quality data to make informed decisions. Our recent report on commercial fishing vessel safety illustrates the need for the Coast Guard to improve the completeness of mission data.

Commercial fishing has one of the highest death rates of any industry in the United States. Vessel disasters are the leading cause of fatalities among fishers. Although the Coast Guard investigates these incidents, we found that rates of accidents, injuries, and fatalities involving commercial fishing vessels cannot be determined. Reliable data are either not maintained or are not collected by the Coast Guard or other Federal agencies.

Having this information could be useful to carrying out the Coast Guard's marine safety mission. The Coast Guard reported that it is taking some actions to address our recommendations, and we will continue to monitor these actions.

Our review of Coast Guard strategic planning illustrates the need for the Coast Guard to also use data on actual asset performance to inform its allocations of assets to field units. We found that the Coast Guard's strategic allocations of assets were based on unrealistic assumptions about the asset performance capacities.

They also did not take into account actual asset condition or maintenance needs. For example, the Coast Guard operates cutters that are 40 to 50 years old. These cutters are hampered by mechanical problems requiring emergency dry dock repairs, which results in reduced availability to carry out their missions.

Officials from one field unit told us that they had planned for 575 hours per vessel per year for one type of cutter in contrast to the headquarters' assumption of 825 hours performance capacity for the same asset. As a result, direction from headquarters, which is based on asset performance capacity rather than actual performance hours, did not provide the field with realistic goals for allocating assets by mission.

Now, our review of Coast Guard performance information illustrates the need for the Coast Guard to also improve the data it uses for establishing its performance goals. For example, the Coast Guard has two performance goals related to its drug interdiction mission. While the two goals capture performance data related to cocaine, they do not capture performance data for any other illegal drugs that the Coast Guard interdicts.

We and others, including the DHS [Department of Homeland Security] inspector general, have reported on the need for measures to accurately assess progress toward achieving desired outcomes.

We recommended that the Coast Guard develop additional goals to address mission activity gaps or explain why certain aspects of mission performance are measured while others are not.

Let me now turn to the issue of transparency.

Our recent reports have identified areas in which the Coast Guard could improve transparency of information used for capital planning. For example, our work on the Coast Guard's annual 5- year Capital Investment Plan found that the plan does not consistently show the effects of tradeoffs that are made as part of the annual budget cycle. This could make it difficult for the Coast Guard to afford its planned program of record.

In 2014 we recommended developing a 20-year plan that identifies all necessary recapitalization efforts and any fiscal resources likely necessary to build these assets. In our recent work we have continued to emphasize the importance of this type of capital planning.

The Coast Guard generally concurred with our recommendations and is in various stages of implementing them. For example, in 2016 the Coast Guard revised its mission needs statement. This statement provides a basic foundation for the long-term investment planning that is to serve as the basis for evaluating the effectiveness of various fleet mixes.

However, the 2016 mission needs statement did not identify specific assets the Coast Guard needs to achieve its missions. It also did not update the annual hours the Coast Guard needs from each asset class to complete its missions.

It is unclear when the 20-year plan will be completed, but its analysis could serve as the foundation for understanding potential tradeoffs that could be made across the Coast Guard's portfolio of acquisitions to better meet mission needs within realistic funding levels. Such an analysis would facilitate a fuller understanding of the affordability challenges facing the Coast Guard.

In closing, our work has found that the Coast Guard can do more in terms of collecting, using, and improving the transparency of information to help meet its mission.

Mr. Chairman, this concludes my statement. I would be happy to take any questions you may have.

Mr. HUNTER. Thank you, Mr. Anderson.

Admiral, let's start here. I want to make it clear so you know what this hearing is about. It is about the implementation of Coast Guard programs. This isn't about whether the Coast Guard does its job or not, because the Coast Guard does its job. It is not about the coastguardsmen not taking risks, because they take risks every day in defense of this Nation. You have 11 statutory missions. We are not talking about that at all.

We are talking about things like this, and we will start with this one. The Coast Guard Authorization Act of 2010 required the following for seamen's shoreside access. Do you know what that means, seamen's shoreside access?

Admiral FAGAN. Yes, Mr. Chairman, I do.

Mr. HUNTER. OK. I will read it for the record here. "Each facility security plan approved under section 70103(c) of title 46, United States Code, shall provide a system for seamen assigned to a vessel at that facility, pilots, and representatives of seamen's welfare and labor organizations to board and depart the vessel through the facility in a timely manner at no cost to the individual." That was 2010.

Every year since 2015, which is 5 years after 2010, the Coast Guard has said that the seafarer access final rule will be issued in the near future. Is it issued?

Admiral FAGAN. No, Mr. Chairman. The final rule is with the administration. And I am happy to provide a detailed brief once it is cleared. But it is in clearance with the administration.

Mr. HUNTER. So I will just add these up. That is 8 years.

Let's go on to the next one. Between 2013 and 2017 Congress appropriated over $2 billion more for acquisitions than the Coast Guard requested in order to meet emerging needs and appropriately position the Service to meet all the mission demands. For the past five budget cycles, that is 10 years, Congress has appropriated $2 billion more for Coast Guard acquisitions than was requested.

Why is that? That is 10 years, five budget cycles—or is that 5 years? Am I thinking terms or years? Five years, sorry, so we will just go 8, there is 5.

Admiral FAGAN. So, sir, I am not sure I completely understand.

Mr. HUNTER. So we—let me phrase it easier. Congress gives you more than you ask for, knowing what you need somehow better than you know, and that is where the GAO metrics come in. And what I am learning this year in the Department of Defense and all the different acquisition changes that we are doing, the metrics that you are using, whether it is you dotted the i's or crossed your t's or did every process point that you were supposed to do, your outcomes are what we want to measure. And we want to measure those, your time outcomes, as well.

Your time metrics are horrible. I mean, if this was an F–35 being built or boats being built, which you went through boats being built with no actual boats being built but spending billions of dollars, that is what this is coming from. There are no metrics for us, for GAO, or for the public to see that you are accomplishing what you are telling us you are going to be accomplishing in the future.

And go ahead and tell us why you don't request the $2 billion for the last 5 years that we have added on to make sure that you can do your jobs.

Admiral FAGAN. So, Mr. Chairman, I would like to talk to how we do utilize metrics. And I talked in my opening remarks about the operational planning process that we use for a repeatable, recurring, predictable process that allows us to make resourcing decisions and resource allocations against the 11 statutory missions. It is a key part of iterating and continuing to approve.

We absolutely engage with the budget process, understand the important oversight role that this committee plays in ensuring that the Coast Guard is properly budgeted and resourced.

Mr. HUNTER. I don't understand. Why is it $2 billion short every year when we give you the $2 billion that you need? And you spend all of it. I mean, you don't say, "we don't need that," and there is $2 billion left over every year, right?

Admiral FAGAN. Yes, sir. And the Commandant has testified and has been very consistent in the need for a stable, recurring $2 billion CIP, 5 percent annualized O&M growth, and the need to begin having stable and recurring money, $300 million a year, to begin to buy back some of the shoreside backlog that the Service currently has.

And we certainly appreciate all of the support that the administration and this committee have given us in ensuring that we are resourced in a manner that helps us continue to meet our mission needs and requirements.

Mr. HUNTER. OK. So we will hit CIP here last. It is always late. And from what I was looking through and reading the materials up to this, it looks like the life-cycle cost estimates are always off. The last one, the last CIP you had for the NSCs [National Security Cutters], didn't even have the new NSCs on it, the ones that were appropriated for that the Senate put in, the extra NSC.

If you could speak to the CIP. Is it the process within the Coast Guard that it is all fouled up so it is not able to be done on time for the President's budget request and for our authorization process for the Coast Guard bill? I mean, what is the problem and what will help?

Admiral FAGAN. So, sir, I acknowledge that the CIP, the fiscal year 2019 CIP is late. We are as anxious to deliver the CIP to the committee and our overseers as you are to receive it. I assure you that it is a critical document to us as it is to you, and we are working with due diligence and a sense of urgency to complete that document.

So the bad news is it is late, I acknowledge that. We are working to be responsive and deliver it to this committee, sir.

The good news is when the CIP is delivered some of the life-cycle cost estimates, particularly with regard to the NSC, will have been updated and revised, and I think you will be pleased to see that progress, sir.

Mr. HUNTER. And hopefully it will be 2018 and not 2016 or 2017. Admiral FAGAN. Yes, sir. I am hopeful that we will deliver the fiscal year 2019 CIP shortly, sir.

Mr. HUNTER. All right.

Mr. Garamendi.

Mr. GARAMENDI. Mr. Chairman, thank you. Your line of questions raises a significant problem for us in that we are given information about what the Coast Guard wants and it is significantly different from what we anticipate the Coast Guard needs, and you were referring to the couple billion dollars annually.

I would like to go into the process a little more deeply here. The budgeting process generally begins with OMB [Office of Management and Budget] sending the various departments its expectations, that is OMB's expectations for the coming year. You then have a pass back and OMB then checks that out, and then that eventually becomes the President's budget.

Is that essentially what happens here?

Admiral FAGAN. Yes, sir.

Mr. GARAMENDI. I thought so. So the question is, is it an answer to the chairman's question that the problem begins with OMB sending the Coast Guard a minimum budget that doesn't reflect what we perceive the needs of the Coast Guard to be? And I am not sure you want to answer that. Let me just make that a statement. My recollection when I was at the Department of the Interior, that is how it works, and so there is a constant tension back and forth.

What we need, therefore, is at the end of our process to have a clarity over the 5-year period as to what is expected, what we expect the Coast Guard to do that is consistent with our authorization. Can the Coast Guard provide that information to us without OMB sidetracking the data?

Admiral FAGAN. We engage in the budget process as you have outlined it, sir, and we are working with a sense of urgency and diligence to deliver that 5-year outlook to the committee. We understand how critical it is for the oversight and investment decisions that need to be made, and we are fully engaged with the budgetary process, sir.

Mr. GARAMENDI. Perhaps a way that we could address this issue is that at the conclusion of our annual or semiannual authorization for the Coast Guard we have authorized a plan, a certain number of ships for this class or that class, certain number of personnel and so forth for the out-years. And perhaps maybe the answer here is for us to require the Coast Guard to report its progress in the authorization that we have provided in the most recent reauthorization of the Coast Guard functions.

Could the Coast Guard do that without OMB screwing up the numbers? I should not add that last clause to it. Let me just say, could the Coast Guard report back on its progress in meeting the authorized activities?

Admiral FAGAN. Sir, we would absolutely be responsive to the desire to understand how the authorizations have been executed.

Mr. GARAMENDI. I think there may be a way of having the Coast Guard be responsive to the authorization, rather than responsive to the annual instructions from OMB. I will just let that hang out there and consider that myself and perhaps with the chairman and staff that that might be way we can be up to date and require that the Coast Guard at least be responsive to what we have authorized to be done, recognizing that we are not the only player in the game. Certainly the President and the administration are as they allocate in their minds the resources of the Nation.

Just generally, Mr. Anderson, it appears as though your report indicates that the Coast Guard is making steps to and progress in answering the performance questions that you, GAO, have raised over the years. Is that correct?

Mr. ANDERSON. Yes, that is correct. And I would like to applaud the Coast Guard for concurring with virtually all of our recommendations on improving data quality in the context of performance goals.

I would also like to note that the Coast Guard routinely develops corrective actions and articulates those corrective actions when a goal is unmet. They put these corrective actions into the APR [Annual Performance Report], and we applaud them for that.

I think our point on the corrective action score in terms of those unmet goals is that some of the corrective actions aren't measurable or time bound, and those are criteria that we would like to see for full closure of those recommendations.

Mr. GARAMENDI. OK. Before the chairman disciplines me, I will yield back.

Mr. HUNTER. I thank the gentleman.

Mr. Weber, you are recognized.

Mr. WEBER. Thank you, Mr. Chairman.

Admiral Fagan, a couple of questions for you.

Does the administration get periodic updates or do they just when the OMB—when you all come in for a request, does the OMB just get it once a year and look at it?

Admiral FAGAN. So, again, sir, we fully engage in the budget process at all of the multiple steps and endeavor to be responsive to each of the pieces in the process as the President's budget is developed.

Mr. WEBER. So is there someone who interacts with the White House kind of on a periodic basis or is it just you walk in there and present a budget proposal to him?

Admiral FAGAN. We have a budget shop within the Coast Guard that engages the Department and the administration and entities as other organizations do as the budget process is developed.

Mr. WEBER. OK. Of course you know about the annual report that says if your stated goals were modified to address "mission activity gaps." Is that addressed periodically or is that something you all report with the GAO, back to GAO, on an annual basis? Or do they monitor?

Maybe that is a question for Mr. Anderson.

Do you monitor that periodically?

Mr. ANDERSON. We monitor that when we have ongoing engagements, looking at that within the scope of our work for Congress. So as that comes up in our ongoing engagements that would be something that would be reportable.

Mr. WEBER. OK.

And, Admiral Fagan, do you agree with that assessment? Do you all have that kind of dialogue periodically?

Admiral FAGAN. I do, with regard to measures in the operational planning process that we utilize. It very much is an ongoing iterative discussion. We work with the Department, with GAO. The goal is to be better at the end of each cycle than we were when we started it.

Mr. WEBER. So they have stated that you have mission activity gaps. Do you—"test" is not the right word—but do you monitor that, do you come back and say, "Here is how you can address that gap and how you can do it better"? Do you all have that kind of interaction on a periodic basis?

Admiral FAGAN. Where there are gaps there is a very healthy discussion of what the gaps are, how to make the measure more reflective of performance in the mission set. And, again, it is very much an ongoing and continuing conversation so that we have the best, most reflective measures of performance available to our overseers and to the public.

Mr. WEBER. Other than not calling you all in here to testify to too many hearings, how can Congress help you all make that goal? In other words, close those gaps. How can we help you do that?

Admiral FAGAN. So, generally, sometimes the gaps are a result of how the performance measure is framed. Others can be a result of not being the only entity that controls an outcome.

Congress has been exceptionally supportive of the Coast Guard and of helping us to ensure that we are best meeting those performance metrics and measures, and we look forward to continuing an open and continuing dialogue with our overseers and with the other governmental agencies that share that same——

Mr. WEBER. And we appreciate that. Vice Admiral Karl Schultz came to my district in Texas, the three coastal counties, after Hurricane Harvey and just did an absolute—the Coast Guard just did an absolute yeoman's job. And I am not even sure that is accurate enough. So we appreciate that.

Would you give us any suggestions, Admiral, as to—and I have got about 1 minute left—how else we could help you?

Admiral FAGAN. Sir, this committee and Congress have been exceptionally supportive. We absolutely are committed to providing the information the committee has asked for to enable the oversight and resourcing conversation that is so critical to the success we achieved in Texas this past summer and the ongoing successes in the mission sets around the country, sir. Thank you.

Mr. WEBER. You betcha.

All right, Mr. Chairman, I am going to yield back.

Mr. HUNTER. I thank the gentleman.

I think what you have done, you have actually beat us down. We know if we require more reports from you, we just won't get them in the future, so we won't ask for any.

Ms. Plaskett is recognized.

Ms. PLASKETT. Thank you, Mr. Chair.

And first I want to just let everyone know how incredibly pleased and grateful I am to be here on this subcommittee and to have my first subcommittee hearing with Admiral Fagan.

Thank you so much for all that you and your men and women of the Coast Guard did for the Virgin Islands. I have been grateful for you even bringing me back home once or twice on some of your vessels as I was trying to make my way back home during the aftermath of Hurricanes Irma and Maria.

So the people of the Virgin Islands are enormously grateful to you all for the work that you have done and continue to do on our islands.

When I was looking at the GAO report one of the things that I noted was a discussion about the use of more realistic asset performance data. And they said that they reported "that the Coast Guard's strategic allocations of assets were based on unrealistic assumptions about the performance capacity of its assets and did not reflect asset condition and unscheduled maintenance."

Can you talk just briefly about what your assessment of that portion of the report means?

Admiral FAGAN. So we continually evaluate what type of performance and effect we would be able to achieve with the assets that we have assigned. And performance measures that we use for the different asset types have served us well. There are some areas GAO has pointed out, and we are looking at how to refine those measures to best reflect how the resources are able to perform in the mission sets.

Ms. PLASKETT. Well, to me, in a best case for the Coast Guard scenario, what this appears to me to mean is that you are trying to outperform yourself on assets and with material that doesn't meet the needs of what you all would like to do, that you are pushing your people and maybe performing in a manner that outperforms antiquated or old, you know, when they talk about unscheduled maintenance and the need for cutters that were 50 years old and the cutters were hampered by mechanical failure requirements.

If, in fact, you had the assets that you needed and were the best case scenario in terms of your budget allocations and the budgets that we have given you, then you would be able to perform more effectively. You know, as a parent that is what I would tell if my kid got a bad grade and they said that about me. What do you think?

Admiral FAGAN. So thank you.

Which is why we have undertaken the significant recapitalization that we have of the Coast Guard. The legacy assets do experience increased maintenance days, and those are days that they are not employed against the mission set.

The quicker we are able to bring the new assets online, I will use, for example, particularly in the Virgin Islands and the Caribbean, the Fast Response Cutters that are being deployed into that region are being deployed for great effect against some of the threats and the mission set and are experiencing some significant success.

Ms. PLASKETT. But as you talk about the fast cutters that are being used, I know, in the Virgin Islands, how is the effectiveness of that in comparison to the use of National Security Cutters or Offshore Patrol Cutters or having a larger amount of cutters to be able to meet the needs of what you see are the national security threats in that region?

Admiral FAGAN. So the new modernized cutters, the National Security Cutters, operational, the OPCs [Offshore Patrol Cutters] and the FRCs [Fast Response Cutters], are incredibly capable assets that with other force packages, aircraft and small boats, increase the effect that we are able to achieve in the mission sets. A National Security Cutter, a day afloat on a National Security Cutter provides a much larger impact than that same day afloat on a legacy 378-foot High Endurance Cutter.

Ms. PLASKETT. Thank you.

And just in closing, do you have any report on the status of the Coast Guard marine debris removal operations? I know there have been recent Washington Post articles about the state of debris removal in the Virgin Islands, and I know a lot of that is the Army Corps of Engineers. But has the Coast Guard been deployed to assist in moving those things directly off the island and are facing regarding these activities within the Virgin Islands and Puerto Rico, as well?

Admiral FAGAN. We were heavily involved under ESF–10 in a number of spill responses and heavily involved in removal of recreational vessels. The larger debris removal I don't have any information on.

Ms. PLASKETT. OK. Thank you. I yield back.

Mr. HUNTER. I thank the gentlelady.

Mr. Graves, you are recognized.

Mr. GRAVES OF LOUISIANA. Thank you, Mr. Chairman.

Admiral, thank you very much for being here.

First issue. In August of 2016 the Coast Guard proposed a final rule for TWIC [Transportation Worker Identification Credential] card readers at certain facilities adjacent to waterways. We had strong concerns that we spoke about in this committee room. We had subsequent followup meetings with the Coast Guard on that. And we are seeing inconsistent guidance coming out of the Coast Guard to some of these facilities.

And just to give you a little background, I think you know this, but you have a number of facilities that may have maritime access or a maritime component, but that can be a very minor component of a much larger facility. Yet you are requiring TWIC card readers at the entire facility.

These can be costly. Some of the information or guidance on securing some of the private information related to this has not been provided to some of these facilities. And I will say it again, that the Coast Guard indicated to us in this room and indicated to us in meetings that this could be delayed and sort of probably some stakeholder engagement, maybe looking at some tweaks to it.

So now we are in a situation where August of this year is implementation. All these facilities have been operating under the assumption for the last several months that there was going to be some type of change or revoking this rule while folks thought through it a little bit more. And now we are getting inconsistent guidance on whether or not this is going to actually apply in 2018 of this year.

Can you give me a little bit of an update on what we should be telling our constituents?

Admiral FAGAN. Yes, thank you. The Coast Guard is working in support of DHS—

Mr. GRAVES OF LOUISIANA. Would you mind going closer to the microphone?

Admiral FAGAN. I am sorry, sir. The Coast Guard is working in support of DHS, who is the regulatory lead on the effort. We are well aware of the confusion that when the final rule was published this summer with regard to extent of applicability, and particularly the concern no opportunity to comment. And so we are working with DHS to ensure that we provide the clarity that has been sought by your constituents.

Mr. GRAVES OF LOUISIANA. So it is March now. We are talking about August. That is 5 months. And some of these potential changes to comply with the rule candidly would probably take 1 year of construction and in some cases maybe even longer.

And I know you are aware we have all these chemical facilities along the Mississippi River between Baton Rouge and New Orleans, one of the largest industrial corridors in the world, and all of them have been operating under the assumption that this was going to be pulled back, because that was the signal that we were all given.

And so we have strong concerns with the certainty and the ability of these facilities to even comply. And, again, we really need to know what to tell them to do at this point. I don't think it is fair, quite frankly, to come in and say, well, no, actually we are going to leave it in place as it was before, because I think we all acknowledge that the ability to comply probably was a little bit too challenging for some facilities based on how they are laid out.

I am really just looking for some guidance on what we are supposed to tell them. And what would you do if you were a plant or some other type of facility adjacent to one of these waterways right now?

Admiral FAGAN. Yes, sir, I understand the need for clarity. Again, we are working within the rulemaking process at DHS diligently to provide better clarity and certainty for the entities impacted by that rulemaking, sir.

Mr. GRAVES OF LOUISIANA. All right. I am going to say it one more time, that you can't comply with it at this point. You can't. Most of these places don't have enough time.

And if this is a problem at DHS or somewhere else I really would appreciate you all helping us identify where the challenge is, because it is just not fair. There was an opportunity for stakeholder input, and it is not fair at this point to come back and say, well, we are just going to leave the previous rule in place and apply it in 5 months.

So I appreciate you taking that back. If you want to talk to us offline about where we should be directing our phone calls or other things, I would appreciate that.

The fiscal year 2019 budget has approximately $720 million in it for three heavies. And I understand that last week it went out for an RFP on design and construction.

Can you talk a little bit about what your expectation is for timeline moving forward on when those would actually be operational and just some type of Gantt chart on what we are looking at moving to the right?

Admiral FAGAN. Yes. On Friday we did release the RFP broadly, looking for resource proposals for beginning construction of a heavy icebreaker. The Gantt chart moving forward is hopeful that the first heavy is delivered 2023 with then out-year delivery of additional hulls.

We used industry studies as a best practice to help in development of those RFPs and as a means to just help accelerate the quality of the design and the timeframe moving forward, and we are standing by and ready to execute when additional moneys are appropriated.

Mr. GRAVES OF LOUISIANA. All right. So first hull 2023. The next two obviously to the right of that. And you think things stay on schedule right now and anticipate funding obviously in the out- years to keep this thing going?

Admiral FAGAN. Yes, sir we have never been closer to having a heavy icebreaker as a Nation. It is critical. My first assignment 35 years ago was on the Polar Star. I am well aware of the criticality and national security need of replacing that vessel, and I am confident we are on a positive timeline, sir.

Mr. GRAVES OF LOUISIANA. They are not paying attention, so I am going to keep talking.

Can I keep going? Is that all right?

[Audio malfunction in hearing room.]

Mr. GRAVES OF LOUISIANA. Oh, come on. You know what, Mr. Garamendi, you all go so far over. I sat here and watched you all like go 20 minutes in the red. So you all just keep talking. You all are good.

Three mediums. The budget—I am sorry, I am going to talk about three mediums then talk about another issue in the budget.

Can you talk about three mediums and sort of just what you expect timeline on that and a little bit about operations and again delivery on something on three mediums?

Admiral FAGAN. So the priority are the heavy icebreakers. We are focused right now on opening the delivery and product line on the heavy icebreakers, and the medium becomes part of those out- year conversations as we bring the first heavy online.

Mr. GRAVES OF LOUISIANA. Am I correct in recalling, though, that you all did identify the requirements for three mediums as well? Admiral FAGAN. Yes, sir, three heavies, three mediums.

Mr. GRAVES OF LOUISIANA. OK. And so you are focusing on the three heavies, and obviously those are going to be significant costs on those. The mediums you are expecting further right. So you are expecting all three heavy hull deliveries, then moving to mediums. Do you expect any concurrent——

Admiral FAGAN. The exact timeframe on the mediums moving forward are very much part of the ongoing discussions. Focus absolutely is on the heavies, and then we will look to leverage what we can from the heavy icebreaker.

Mr. GRAVES OF LOUISIANA. OK. I think I can speak for Mr. Hunter, Mr. Garamendi on this, I am not sure that Congresswoman Plaskett cares about icebreakers in the Virgin Islands maybe. But as you know, the comparison of U.S. capabilities to other Arctic nations, we are significantly further behind where we are. You have a number of advocates here to ensure that we have the capabilities we need to address many missions of this Government. And just want to urge you to keep that on the front burner.

Last issue. The Waterways Commerce Cutter, I think there is $5 million in fiscal year 2019 budget for that. Just curious if you can just very quickly again give us a summary of what you expect that to look like moving to the right.

Admiral FAGAN. Yes, sir, there is money in the budget for the Waterways Commerce Cutter. Again primary focus with regard to the next major project is the heavy icebreaker, but very much beginning requirements development and the work that marches us forward with regard to acquisition and replacement in the Waterways Commerce Cutter. That timeline, as well, will resolve itself, but again, a sense of urgency again and another major recapitalization need.

Mr. GRAVES OF LOUISIANA. Great. Thank you.

Mr. Chairman, I would like to yield back all of my remaining time to Mr. Garamendi.

Mr. HUNTER. You owe us now.

Mr. Garamendi, you are recognized for further statements or questions.

Mr. GARAMENDI. Given the extraordinary response of the Coast Guard to the hurricanes, I would expect that this year's metrics are not going to meet what was anticipated. Unintended or unanticipated events will undoubtedly modify those metrics.

And so in the reporting to us and to the GAO how can this be taken into account, by simply noting that we had 2,000 personnel that were doing things other than chasing drugs in the Caribbean? They were chasing survivors. Does that work? Does that fit for the GAO and how you might respond?

Admiral FAGAN. So you very accurately identified the opportunity cost to the organization associated with the responses this summer. How those are accounted for and the impacts on the performance metrics and measures will be part of the operational planning process in our internal look, the repeatable, reliable process.

I am not an expert on how we will do that bookkeeping, but I would expect that there will be some adjustment that reflects the significant level of effort expended by boats and assets and cutters against the disasters this past summer.

Mr. GARAMENDI. And I suppose GAO, as you do your reviews, you take these things into account.

Mr. ANDERSON. Absolutely. Our October 2017 report did focus on the goals and the reasons for unmet goals and highlighted seven case studies where we took a look at performance goals that crossed a variety of missions. And we analyzed some of the meaning behind why goals weren't met.

We would just like to point out, though, that your question does underscore something very important about the handful of reports that we have used to support this testimony statement, and that is, good data is needed for good allocation of assets. And we pointed out several times where there is a bit of a mismatch between the actual use and the assumed use of a handful of assets, which could limit the ability of the Coast Guard to surge.

So bringing this back to our central message, better data available on actual asset use will provide a clearer picture for the Coast Guard to then reallocate when there are unanticipated events.

Mr. GARAMENDI. OK. Thank you.

One of the results of the hurricanes was a significant loss of property assets, Coast Guard assets, and we have approved a budget for the replacement of those assets, or the replacement improvements, and whatever other maintenance might be required, to the tune I think of something less than $1 billion.

Here is where I want to go with this question, and I want to use this for the other programs of record that are out there, and that is that it is important to me and I believe to this committee that we have a continuing update not every month, perhaps not every other month, but maybe on a 6-month cycle on the progress of these measured programs.

Mr. Graves spoke at length on things that are really not his turf, that is our turf, my turf, the chairman's turf, we call that our icebreakers, not Graves' icebreakers, but we will get over that, and we will deal with that personally here.

But he raises a very, very important point: What is the progress? How is it going? What problems have you encountered in the design, the money, the contracting, and the like? And then carrying on to the other light icebreakers, other heavies and other lights. We need to know that.

And so for all of the major programs that you have underway, the Offshore Patrol Cutters, the other various capital programs that you have, and now the followup on the hurricane repairs, these are important things for us to keep track of for two reasons. One, we may want to modify the program of record. Or we may want to cancel or augment. And we can't do that unless we know where you are in the process.

Now, I understand you do this occasionally, but not in a periodic way, so that, say, every 6 months where are you with the I think it is Fast Response Cutters that are built in Mr. Graves' district? Where are you with the Offshore Patrol Cutter? All of those things.

Where are you periodically along the line so that we can say, oh, my, there is a problem that comes about for whatever reason, and then we can address it or not, or encourage you to address it? Is that a sensible way for us to keep track of where we are in these programs that we have laid out for the Coast Guard to do? And can you do it?

Admiral FAGAN. We very much do what you have described with regard to program management and oversight of our major acquisitions, any one of which you have named. It is a regular, recurring, senior leadership-driven review. We are responsible to DHS as well.

And so the process is there. And to the extent that we can and should share information with our overseers I am confident we have got a well-developed ongoing process with regard to ensuring that we are responsibly and transparently and reliably spending the resources that are generously allocated to us.

Mr. GARAMENDI. So internally it is being done on these major programs.

Admiral FAGAN. Yes, sir.

Mr. GARAMENDI. And I suppose on the other, the 11 specific tasks that you are required to carry out, search rescue, so on and so forth, drug interdiction, and the like.

I am going work with the chairman and see if there is some way that we could receive just a periodic review of that.

Much of the GAO report appears to me to be one of setting up systems of metrics so the Coast Guard knows where it is on all of these tasks that it has to do, that there is a methodology of reporting, so many sailors doing—excuse me, so many coasties doing whatever they are supposed to be doing, so much drug interdiction, so much work on boating accidents and so forth. And that seems to me to be useful internally for the Coast Guard to keep track of where it is and what it is supposed to be doing.

For us, we have a somewhat different task. We will be looking at the larger picture, reviewing particularly where the big money is going, where the big tasks are going, and if we can set up some sort of a repetitious, not every month, maybe every 6 months, but certainly no less than every year, where we are reading, OK, where are you on the icebreakers, which probably would be like once every other month because of the way in which it is now gearing up.

Anyway, that is kind of where I am coming from, and I would like to see if there is some way we can keep track of that.

Right now it is hit and miss depending upon when we are able to call a hearing, and maybe we are asking for it and maybe we are not. But just as a way of keeping track of where we are on these major programs would be useful as I look at what the Coast Guard is responsible to do.

With regard to the annual budgeting and so forth, I understand OMB, and I understand all the games that you have to play there, and I understand that you really don't control your budget, and that the Department of Homeland Security is probably ripping your budget off to build something, maybe a wall, I don't know. But whatever. We understand you don't control all of that. We can, if we have knowledge and information.

So thank you. I will let it hang out there and see.

Mr. HUNTER. I thank the gentleman.

That is a great question, if you could track the icebreaker progress. But if you can't do the CIP and everything else is late and your internal metrics and targets don't match your external outside-in metrics, I think it is a futile thing to ask you to report to us because you will just be late and say: We are working on it, it is important to us and we are working on it. But we can talk about that.

Mr. Anderson, let's go here, because we are talking about DHS. Is it DHS' prohibition on the Coast Guard that does not allow them to release a full annual performance report?

Mr. ANDERSON. I believe that would be a better question for the Coast Guard witness.

Admiral FAGAN. So, Mr. Chairman, with previous negotiations with DHS with regard to releasing the full report, the preference was for partial reporting within the totality of the DHS report.

We have every intent this year and are very close to publishing the full performance measures report. I am confident that here quite soon that will be publicly available, the total report will be fully published.

Mr. HUNTER. Does the Coast Guard feel that it is being asked to record too many metrics, too many things that don't have anything to do with the outcome of your 11 missions?

Admiral FAGAN. The performance measures report and the process that we use to look at those measures and evaluate our performance and effectiveness are part of the process, the annualized, repeatable process, and are absolutely critical and essential to us continually evaluating how we are performing and how——

Mr. HUNTER. But it can't be that essential, because a lot of these reports are late.

Admiral FAGAN. So are you referring to the performance report, sir?

Mr. HUNTER. No, CIP. But I am just saying these are all things that are due, these are all things that inform you about your progress and what you need and how much money you need and how well the Coast Guard is doing. If they are not on time, like the CIP, how do you use that to inform yourselves?

Admiral FAGAN. So, sir, we are absolutely committed to delivering a CIP. I acknowledge that we are late on the deliverable. But that document is as valuable to us as it is to you, and we are committed to getting it to the committee as soon as possible, sir.

Mr. HUNTER. Let me ask you this, Mr. Anderson. How do the Service's 11 statutory missions align with the DHS goals for the Coast Guard?

Mr. ANDERSON. Well, in our recent review, we found there were 37 performance goals that are meant to align with the 11 statutory missions. Those goals either fully or partially aligned with the statutory missions.

Where there wasn't perfect alignment, we did recommend that the Coast Guard identify new goals needed or, alternatively, explain to Congress and to others why these proxy goals fit the bill.

One example here is on drug interdiction. The Coast Guard uses cocaine as a proxy. That doesn't reflect the entire suite of drug interdiction activities or all the narcotics that the Coast Guard ultimately interdicts. So we asked the Coast Guard——

Mr. HUNTER. Wait. What you are saying is they have metrics for coke, but not marijuana or heroin or whatever?

Mr. ANDERSON. Reportable metrics. The Coast Guard does—— Mr. HUNTER. They obviously have metrics on how much weed they interdict every year, right?

Mr. ANDERSON. Yes.

Mr. HUNTER. They just don't give that to you?

Mr. ANDERSON. They don't give that to you.

Mr. HUNTER. Got you. Why is that?

Mr. ANDERSON. That would also be a better question for the Coast Guard.

Mr. HUNTER. OK. Admiral, why would it affect—is this just an oversight or——

Admiral FAGAN. So we do use—cocaine, volume cocaine interdiction is one of our key metrics within the counternarcotics mission set. We acknowledge that there are——

Mr. HUNTER. Is that because it is the most expensive drug out there or most prevalent on the high seas or what?

Admiral FAGAN. It is the metric we are currently using, sir. We acknowledge that there are other indicators that would perhaps be better indicators of performance in the mission set, and we are very much—we look forward to having that conversation with GAO.

Mr. HUNTER. But you keep track of that.

Admiral FAGAN. We do track interdictions, detainees, yes, sir. Mr. HUNTER. You just don't—OK, so just help clear this up. Why don't you give it to us?

Admiral FAGAN. I guess I was not aware that we weren't sharing that information.

Mr. HUNTER. Have we not requested it?

Admiral FAGAN. We do track that. We do track that, sir.

Mr. HUNTER. OK, thank you.

That is really all I have unless Mr. Garamendi or the other members here.

Ms. Plaskett, you are recognized.

Ms. PLASKETT. Sure. Thank you.

Mr. HUNTER. We are always happy to have somebody else chime in besides me and John.

Ms. PLASKETT. Anytime.

I just wanted to ask a couple of questions regarding commercial fishing classification. The classification standards that apply to virtually every other class of commercial vessels, in general, these standards have improved greatly the overall safety record of commercial vessels.

Mr. Anderson, does the GAO agree that classified standards improve the safety of the commercial vessels?

Mr. ANDERSON. Well, that is an area where we don't believe data exists. When we performed our performance evaluation, one of the key takeaways was that the Coast Guard needs data to determine whether classing has a positive effect on safety. So this is where there is a paucity of data.

Ms. PLASKETT. So it may not. It may or may not. You are still looking for the data to make that determination?

Mr. ANDERSON. Yes. We have ultimately recommended that a working group be formed to go get data such as that, because we believe that first we need the data to determine whether or not classing has a positive effect on safety.

Ms. PLASKETT. And why don't you have the data at this point?

Mr. ANDERSON. Well, there are various entities that ultimately collect the data. Coast Guard collects some. Other Federal agencies collect others. State regulators also collect some.

Hence, our recommendation for a working group to bring together those parties, to identify data on the number of commercial fishing vessels, which basically puts the Coast Guard in the position of having a denominator. Right now it essentially has a numerator.

Ms. PLASKETT. So in your recommendation for this working group, you mention that the Coast Guard, NIOSH [National Institute for Occupational Safety and Health], National Marine Fisheries Service should form the working group. Do you think that other stakeholders, such as State organizations or industry representatives, should be in that working group as well?

Mr. ANDERSON. Yes. We do believe that others should be in that working group, in large part, for the reasons I stated, that some of those entities have data on commercial fishing that may not currently be available to the Federal agencies.

Ms. PLASKETT. OK. And it is your recommendation as well that the Coast Guard issue regulations or guidance to clarify or implement alternative safety compliance programs, which you referred to as alternative to class approach. Is it necessary to issue formal regulations or would guidance be sufficient in this area?

Mr. ANDERSON. Well, I think in this area there are some advantages to regulations. Regulations set up a formal process. Regulations also invite a number of different perspectives to the table— including industry representatives. And it is time-bound. So I can see some advantages to the regulations route.

Ms. PLASKETT. Now, were you also able to determine whether the regulations or guidance was helpful in less than commercial class vessels that the Coast Guard reviews?

Mr. ANDERSON. I don't believe that was within the scope of that engagement.

Ms. PLASKETT. Have you done it in the past, do you know?

Mr. ANDERSON. I would have to get back in touch with you with an answer for that one. I can talk to my subject matter experts back at GAO.

Ms. PLASKETT. OK.

Admiral Fagan, do you know if that has been done, reviewed in the past, the effectiveness and the efficacy of the regulations to safety, of the Coast Guard having regulations or guidance on other vessels that are not commercial vessels?

So I am thinking in the Virgin Islands, we don't have large commercial vessels that one thinks of specifically, but we do have smaller vessels that are constantly around the island, whether they are yachts or individuals going on excursions or even private boaters in the area.

Do you know if there are studies that determine the effectiveness of those guidelines versus regulations?

Admiral FAGAN. I am not aware of a study that looked specifically at guidelines versus regulations. We, as we look at different types of vessels and the risk, use regulations and policy and voluntary safety systems, a whole host of tools, to help bring better safety records and safety for the people that are operating those vessels.

Ms. PLASKETT. Because I know, while I don't mind and I see the benefit to all coming on a lot of the vessels in the Virgin Islands for security reasons as well, not just safety of the vessel, I know that our weekend boaters, going to Buck Island in particular, you know, might have a different view of that.

So having the data I think would be really important. And I think that GAO, have they looked into the implementation of the classification requirements, whether or not which ones are necessary and which ones are most effective, if that has been put in place as well?

Mr. ANDERSON. The scope of that report was really on the need for more reliable data in these commercial fishing vessels.

With your permission, I do have one of my experts here to assist, if you would like an answer to that question specifically from one of my colleagues.

Ms. PLASKETT. If you could give that to us afterwards, that would be helpful.

OK. Thank you very much.

Mr. HUNTER. Mr. Garamendi, you are recognized.

Mr. GARAMENDI. I think I will do this for the record, just to save some time here, to follow up on what Ms. Plaskett was talking about, and it is the fishing classification standards. We have that for the larger commercial vessels. They are in line with the IMO [International Maritime Organization] standards. I believe that we have exempted the smaller boats, smaller fishing boats.

And the question is, from the experience of the Coast Guard, would we improve the life safety if the smaller boats were not exempted and had to meet higher standards or perhaps the international commercial standards?

Just we need to get into that. That is an issue that comes back here on a regular basis. And it would be helpful to have some of the Coast Guard experience and advice on that issue.

So I will let that hang out there. I doubt that you will have an answer right now, but we will give it to you for the record.

> [The information from the U.S. Coast Guard follows:]
>
> Operators of all vessels, both commercial and recreational, benefit from enhanced safety standards and practices. Preventative minimum safety is the basis of meeting the standards and regulations presently available, whether at the industry, Federal Government, or international convention (treaty) levels. For commercial vessels, the standards and regulations are based on the vessel's operations, the size of vessel, the amounts and types of cargo or passengers carried, and where the vessel operates. The standards and regulations for commercial vessel operations and construction are based on the vessel's intended use and the types of waters in which the vessel will operate.

Mr. GARAMENDI [continuing]. Also, the chairman and I were just trying to figure out how we can stay on top of the activities, the major programs that the Coast Guard is responsible to carry out. There are 11 of them. Then there are all these new ships that you are building, cutters that you are building, aircraft that you are repurposing, the C–27s and the like. And you, obviously, have heard frustration about the CIP and the lateness of it.

In a question that I asked a while ago, Admiral, you indicated that you do quarterly reviews, or maybe more often than quarterly, probably 6 months also, updates on how these major programs are going in the internal management of the Coast Guard. And I assume that is all the way up to the Commandant, sit around a table and say, what is happening with the C–27s out in McClellan. Are we getting them done? Are we not getting them done? What is the problem?

To ameliorate the frustration that we have about the CIP, which seems to be not timely, presented to us in a timely way, I think I am going to suggest that we pursue a more informal but more often review of where we are on the major programs that we follow.

There is a lot of money involved. They are critical to the future work or the ability of the Coast Guard to carry out its missions. And if you have that internal information, I don't know if it needs to be edited, but it would surely be helpful to us to keep track of where we are as we do our work on authorization and appropriations. We are not the appropriating committee, but we do talk to them from time to time, and they should see things our way, and when they don't we try to persuade them otherwise. But, taken together, I think we would be able to keep better track of where the Coast Guard is on its major programs and activities and simultaneously be able to be of more immediate help when needed on whatever the problem might be.

So I am going to let that hang out there for your review and someday over a cup of coffee maybe we can pursue that. Do you think that would be helpful, Mr. Anderson?

Mr. ANDERSON. I am sorry. Could you specify the part about which part would be helpful?

Mr. GARAMENDI. That it would be helpful to receive a, I don't know, let's say a quarterly or at least a semiannual update on where the major programs are. The Coast Guard apparently does it internally.

Mr. ANDERSON. I am aware that we provide a similar service for other major programs and other departments where we do speak to the appropriators and authorizers on a routine basis, such as quarterly.

[Nathan Anderson of the U.S. Government Accountability Office added the following post-hearing supplement to his remark:]

We also provide information on Coast Guard programs in our annual assessment of major DHS acquisition programs.

Mr. GARAMENDI. I thought you might. And it would be helpful, I think, for us in our oversight review. And we do—what, every other year, or every year, every other year—a reauthorization. Every other year a reauthorization. And the appropriations are annual. And so if we can be helpful with the appropriators, this kind of information is important. Maybe less formal, but maybe we just sit on it with a cup of coffee and review where we are. With that, Mr. Chairman, I yield back. Thank you very much. Mr. HUNTER. I thank the ranking member. Let me add, too, I was looking at your program baseline, the approved program baseline, which I assume informs the CIP, right, Admiral? So that your program baselines, which I imagine a bunch of bars on charts, informs your capital investment for your 5-year building plan, basically. But your baseline doesn't have C–130J, C–27, C– 144s, the ninth NSC that we gave, or the six FRCs needed to replace the CENTCOM [U.S. Central Command] FRCs.

If that is needed to inform your CIP, we might even be happy just seeing the program baselines that you have right now until the CIP is completed, whenever that may be. I am not enthusiastic about seeing anything, frankly.

On the Armed Services Committee we do mean things, like take away Suburbans and airplanes from four-stars when they don't get us the things that we require.

We are going to make sure too that we are not overburdening you with reports. Otherwise, that is our fault. Last authorization, we gave you a whole section with the reports that we thought were important. We changed the structure, right?

So now you have one section. We are going to go through that and look at those to make sure that you are not being overburdened and you are only reporting to us what you need to so that those few things could be on time for once, ever.

Lastly, with performance tracking, Mr. Anderson talked about how the higher headquarters says 800-something hours, the guys actually driving the boats say 500 hours, and that leaves you with a 300-hours-something gap. That doesn't help anybody, it really doesn't. And I am just not sure why it is done that way. It is not done that way for any other service that I have seen. If your aircraft is this many years old and it has this many hours on it, here are how many years or hours we expect out of it, right? That is how it is normally done. It is not a top-level thing where they say, this is what it is supposed to do if it was brandnew, and the people on the ground say, well, this is what it can actually do. Am I getting that right, Mr. Anderson?

Mr. ANDERSON. Yes, you are. That report from about 2 years ago that we issued did say that there was a difference between what the field units knew could be used for a particular asset and the direction it was getting, the strategic intent from headquarters. There was a delta there. For the example that you are citing, it was 575 hours versus 825 hours for a cutter.

Mr. HUNTER. I think in the end, Admiral, it looks like the Coast Guard is making things harder on itself by not using consistent metrics, by not aligning with DHS, and not aligning with us on this committee so we can say, hey, your outcomes were great. Because if you look at your missions and you really delve into them and you don't do the 800 hours versus 500 hours, but you look at your outcomes, right, those things that we don't get all of that GAO was talking about that we would like to start getting, if you look at those you are successful.

So it looks to me like it is a problem between point A and point Z. You are getting to point Z and doing a great job. Point A sounds good. But everything in between, it is not matching up. And it is providing us with trouble, and it is providing you with trouble, and it is providing GAO with trouble, and probably DHS with trouble too.

So we will look through this and see what we can do to assist you to cut down on the reports you have to do that aren't necessary and to require the reports that we need. There are only a few of them, because we are not going to tell you how to tend buoys. We are not going to tell you how to jump out of a helicopter and rescue somebody.

Mr. GARAMENDI. Oh, come on. You can do that.

Mr. HUNTER. Right, we would like to. Marines don't do that; we just kill people.

But what we can ask you to do and what we need is a Capital Investment Plan, so we can say, here is how much money you are going to need, so we can go fight for it on your behalf.

And we can't do that the way that we have been doing it, which is with our secret handshakes with OMB and the administration, figuring out what you really need and then adding that to what you ask for, right.

It is not going to go that way forever, because you might have different personalities sitting here and a different personality sitting there, or vice versa, where the handshake control thing just doesn't work.

So, with that, I am not sure we—I mean, we accomplished a lot of venting here, and that is about it, I think.

So thank you very much for being here.

With that, we are adjourned.

[Whereupon, at 3:18 p.m., the subcommittee was adjourned.]

Statement of the Honorable John Garamendi Subcommittee on Coast Guard and Maritime Transportation "Implementation of Coast Guard Programs"

March 7, 2018

Thank you, and good morning. I want to thank Chairman_ Hunter for scheduling this afternoon's hearing to assess the status of the Coast Guard's capital planning and the Service's overall operational performance.

It has been said often that the United States Coast Guard — at all times a military service and also our Federal maritime law enforcement authority— is indispensable to ensuring the safe, efficient and reliable maritime commerce of the United States.

The Coast Guard's eleven statutory missions and diverse statutory authorities enable this multi-faceted Service to function as first responders, fisheries enforcement agents, port state control inspectors, or maritime border security, to name only a few of the critical assignments the men and women of the Coast Guard themselves day in, and day out.

Yet because the Coast Guard remains by far the smallest of the five armed military services, it does all of this with much less personnel and far fewer resources.

This reality puts a premium on the ability of the Coast Guard to effectively budget and plan for the capital resources it needs — its ships, aircraft, shore infrastructure, aids to navigation, telecommunications, and much, much more.

The Coast Guard has taken great strides over the past nine years to improve the transparency and accountability in the development of its five-year Capital Investment Plan, or UP, and I commend the Service for this effort.

I will be interested to hear Admiral Fagan describe the highlights in the latest CIP, and most important, how the objectives and timetables in this CEP stack up against the reality of projected Coast Guard capital budgets and the Coast Guard's continued reliance on legacy assets that are well past their ideal lifespans.

Additionally, because there is always more for the Coast Guard to do, than resources and people available to the Coast Guard to get the job done, it is imperative that the Coast Guard develop thoughtful operational goals and set realistic performance measures to maximize the effective allocation of assets on any given day.

The Coast Guard simply has not the luxury of being able to waste time, effort or resources.

To best husband its limited assets, the Coast Guard needs measurable and accurate performance metrics. Moreover, the Coast Guard needs effective feedback loops to better translate its performance results into operational adjustments to improve performance.

I look forward also to hearing from Admiral Fagan on this topic, and additionally, I will look forward to hearing from Mr. Nathan Anderson from the Government Accountability Office, to better understand GAO's recommendations for how the Coast Guard might improve its operational proficiency.

Thank you.

TESTIMONY OF REAR ADMIRAL LINDA L. FAGAN DEPUTY COMMANDANT FOR OPERATIONS POLICY AND CAPABILITY ON "IMPLEMENTATION OF COAST GUARD PROGRAMS" BEFORE THE HOUSE COAST GUARD AND MARITIME TRANSPORTATION SUBCOMMITTEE

March 7, 2018

Introduction

Good morning Mr. Chairman and distinguished Members of the Subcommittee. It is my pleasure to be here today to discuss the implementation of key Coast Guard programs, most notably those that have recently undergone a Government Accountability Office evaluation.

First, let me thank you for the outstanding support this subcommittee has given the Coast Guard, especially as it relates to the supplemental funding for hurricane response activities. This critical infusion allows the Service to not only rebuild damaged and destroyed facilities, but also provides the ability to build to modern resiliency standards and ensure the best chance of withstanding future disasters. The Coast Guard is an integral part of the communities these facilities serve, and is committed to continuing to execute its missions in these areas. Your support is helping make this possible.

The U.S. Coast Guard is the world's premier military, multi-mission, maritime service responsible for the safety, security and stewardship of U.S. waters and hundreds of miles seaward. At all times, a military service and branch of the U.S. Armed Forces, a federal law enforcement agency, a regulatory body, a first responder, and a member of the U.S. Intelligence Community, the Coast Guard stands the watch and serves a nation whose economic prosperity and national security are inextricably linked to broad maritime interests.

The Coast Guard plays a critical role in maritime border security, by protecting and defending more than 100,000 miles of U.S. coastline and inland waterways comprising the Maritime Transportation System, saving thousands of lives per year, and safeguarding the world's largest Exclusive Economic Zone (EEZ), encompassing 4.5 million square miles of ocean. Securing the Nation's maritime borders and reducing risk to its ports and infrastructure requires a layered, multifaceted approach.

Because of the Coast Guard's unique authorities, capabilities, competencies, and partnerships, the Service is well-positioned to undertake such an approach and meet a broad range of maritime border security requirements. This layered approach allows the Coast Guard to detect, deter, and counter threats as early and as far from the U.S. shores as possible. Indeed, the Coast Guard is fully engaged in answering the call and balancing a multitude of dynamic maritime risks facing our Nation.

Strategic Focus and Resource Allocation

The Coast Guard safeguards the Nation's maritime interests in its territorial waters, on the high seas, and abroad. To ensure alignment with national strategies and best position the Service to address myriad complex challenges, the Service developed Western Hemisphere, Arctic, energy, and cyber strategies. Each year, the Coast Guard conducts a strategic review as part of the annual program plan development process to determine areas of focus for improved performance, what initiatives are needed to better attain operational goals, and what effect they may have on the aggressive, but achievable, targets promulgated for each five-year planning horizon. By using these strategies and plans as guideposts, leveraging the intelligence community, and employing a risk-based approach to direct resources where they are most needed, the Coast Guard is able to mitigate maritime threats with greater precision and effect.

This strategic allocation of resources, bearing in mind specific asset capabilities and capacities, is critical to the Coast Guard's ability to operate in dynamic, vast, and diverse areas of responsibility. The Coast Guard has adopted the Standard Operational Planning Process (SOPP) as a deliberate planning system to communicate strategic intent and influence mission execution based on program planning. The system, after considering all of the competing demands for Coast Guard resources, operational risk, and platform availability, allows Coast Guard operational commanders to designate mission priorities and distribute available asset resource hours optimally across Coast Guard missions.

Delegating the tactical control of assets to operational commanders is central to the Coast Guard's success, providing the Service with the capability and flexibility to perform day-to-day operations while also remaining ready to respond to major national contingencies. This approach aligns resources with strategic priorities while allowing operational commanders sufficient flexibility to manage risk, leverage partnerships, and make trade-offs in responsible ways that make sense for their areas of operations. Feedback loops are integrated into the system, and the Coast Guard has established asset allocation processes to ensure mission accomplishment and metrics to measure outcomes.

The Coast Guard's responses to Hurricanes Harvey, Maria, Irma, and Nate demonstrated the value of its versatility and sense of on-scene initiative, as it mobilized over 2,000 active duty members, 800 reservists, and 150 civilians to directly assist more than 11,000 people, reopen critical waterways, and mitigate damage to the environment.

In addition to this impressive hurricane response, in fiscal year 2017, the over 40,000 active duty members, 7,000 reservists, 9,400 civilian employees, and 25,000 volunteer members of the Coast Guard made great achievements in service to the Nation.

The Coast Guard prosecuted over 16,000 search and rescue cases, saving 4,228 lives, assisting 22,000 more, and preserving more than $76 million in property from loss. Coast Guard men and women also responded to 12,270 pollution incident reports, mitigating damage to sensitive natural resources.

Frontline Coast Guard operators and assets completed over 5,300 security inspections of maritime facilities, 4,468 small vessel security boardings, and 635 boardings of "high-interest" vessels. At the same time, Coast Guard Patrol Forces Southwest Asia, consisting of six patrol boats and 250 personnel, provided near-continuous presence in the Arabian Gulf supporting maritime security operations while providing, critical training and capacity building in the region.

Coast Guard marine safety personnel completed 5,300 security-related inspections at Maritime Transportation Security Act regulated facilities, while conducting approximately 39,000 vessel and 16,000 facility inspections. Concurrently, they added more than 5,800 towing vessels to the inspected fleet as the result of requirements set out in the Coast Guard and Maritime Transportation Act of 2004.

The Coast Guard boarded 5,518 domestic fishing vessels to ensure critical fish stocks remain available for future generations. The Service also patrolled the largest Exclusive Economic Zone in the world, supporting the Nation's sovereign interests and suppressing illegal foreign fishing.

The Coast Guard supports more than $4.6 trillion of economic activity vital to our Nation's economic well-being, and it does so by facilitating the movement of people and goods through the Nation's waterways. The Coast Guard maintains the world's largest aids to navigation system, servicing nearly 30,000 aids and responding to over 7,700 discrepancies while reviewing 3,600 marine event permit applications and completing over 10,000 waterways activities. The Coast Guard identified and tracked more than 1,000 icebergs threatening ship traffic in the North Atlantic, and dedicated 5,300 operational hours to enable movement of commerce worth over $1.5 billion through ice impeded waters of the Great Lakes and Eastern Seaboard."

Strategic Challenges

Maritime Border Security

The Coast Guard is pivotal to securing our Nation's maritime border. Persistent threats include illegal migration, human trafficking, and the flow of illicit drugs. Transnational Criminal Organizations (TC0s) use networks to commit these dangerous crimes. TCO networks are driven by immense profits from drug trafficking and other illicit activity, and their indiscriminate use of violence weakens regional governments in Central America, stymies legitimate economic activity and development, terrorizes peaceful citizens, and fuels migrant flows.

Coast Guard patrol assets in the maritime approaches to the border get results by employing timely intelligence from an expanding network of partners. The new National Security Cutters (NSCs), Fast Response Cutters (FRCS), and legacy cutter and aircraft fleets achieved impressive operational successes in fiscal year 2017. Coast Guard crews interdicted 2,512 undocumented migrants, and seized a record 223 metric tons of cocaine, 31,190 pounds of marijuana, 6 kilograms of heroin and other opiates, and 168 kilograms of methamphetamines, worth $6.6 billion wholesale. Despite this impressive performance, the Service missed its target for cocaine removal by 3 percent. Critical acquisitions like the Offshore Patrol Cutter (OPC), a more capable and reliable replacement for our outdated Medium Endurance Cutters (MEC), are essential to long-term success in securing America's maritime borders.

Polar Regions

As sea lanes in the Arctic open and access to natural resources increases, the Coast Guard is promoting safe and responsible activity within this vital region. In concert with partner nations in the Arctic, the Service continues to solidify maritime governance regimes, strengthen prevention and response capabilities and capacities, and increase domain awareness in this vast and rapidly changing region.

The Coast Guard's enduring success in the polar regions hinges upon assured access and support for U.S. sovereign interests. The Nation's credibility and ability to protect its interests cannot be guaranteed with an aging national icebreaker fleet and it must be recapitalized. The support of the Administration and Congress to procure new heavy polar icebreakers is critical for U.S. national security.

Cyber Domain

Cybersecurity is not only a vital component of economic and national security, but it is also critical to the Coast Guard's ability to fulfill its statutory responsibilities. In addition to safeguarding its own networks from malicious cyber activity, the Coast Guard is developing regulatory approaches and guidance for industry that will instill stronger cybersecurity protocols in the Nation's critical Maritime Transportation System infrastructure. For example, Coast Guard-led Area Maritime Security Committees (AMSC) are incorporating cybersecurity subcommittees into their current practices.

Success is also dependent upon building, developing, and retaining an appropriate cyber skill set in the Service's workforce. Given the growing global demand for cyber professionals, the Coast Guard will be challenged to compete for this highly specialized and uniquely qualified staff.

Maritime Safety, Security, and Stewardship

The prosperity of the Nation is inextricably linked to a safe and efficient Maritime Transportation System. Increased recreational and commercial waterway usage, compounded by growing domestic natural gas extraction and transport, deeper U.S. ports, and the expansion of the Panama and Suez Canals, has placed growing pressure on the Maritime Transportation System. Continued uncertainty and volatility in the domestic energy sector requires dynamic planning efforts on the part of the Coast Guard and its partners to effectively regulate existing and emerging technologies and ensure the safety, security, and environmental stewardship of the Nation's waterways.

Living Marine Resources

Fishery resources are a critical source of sustenance in a world with an ever-increasing population. The Coast Guard's role in the enforcement of laws and regulations that preserve living marine resources is a fundamental and significant mission. It involves two interwoven missions: preventing illegal catch from over-depleting fish stocks, and providing for the safety of the mariners and vessels who fish U.S. waters and perform one of the Nation's most hazardous occupations.

Like maritime border security, significant assets are needed to ensure compliance with fishery laws. Since the majority of fishery activities are beyond the operational range of boats or even smaller cutters, programs like the OPC are critical to providing the appropriate platforms to detect and board vessels and verify compliance. The dispersed and far-reaching nature of this mission resulted in the Coast Guard achieving only 23 percent of its fiscal year 2017 target number of compliance boardings for fishery enforcement.

In addition to protecting the fish stocks, the Coast Guard is charged with protecting the mariners that harvest them. Over the years, the Coast Guard has increased its oversight of fishing vessels through new authorities delegated by Congress. Currently, the Service is seeking ways to improve the range and quality of data it maintains on active fishing vessels and their key characteristics. Additionally, it is exploring alternatives to address new mandates, such as those requiring certain fishing vessels to be classed. The Coast Guard looks forward to working with Congress as it strives to meet new requirements and facilitate legislative intent in this area.

Measuring Success

As the Coast Guard pursues its strategic goals across all missions, aggressive and achievable indicators of mission success are more critical than ever.. The Coast Guard was one of the first agencies to adopt the requirements of the Government Performance and Results Act when it was first enacted in 1993, and the Service is committed to continued refinement and expansion of its portfolio of strategic and management metrics to ensure that benchmarks for performance are clearly defined. These metrics are developed in consultation with the Department of Homeland Security and Office of Management and Budget, and they undergo a rigorous vetting process. Only some activities the Coast Guard performs are represented by strategic or management metrics. These key measures are complemented by an extensive array of operational measures that assist Coast Guard leadership at all levels in making informed decisions related to the utilization of assets and personnel across the entire spectrum of Coast Guard mission implementation.

Annually, Coast Guard strategic measures are reported in the Department of Homeland Security's Annual Performance Report and made available to the public. Additionally, the Coast Guard prepares its own Annual Performance Report, which contains the results of all strategic, management, and key operational measures. In alignment with a recent GAO audit recommendation, the Coast Guard will make the Annual Performance Report for fiscal year 2017 publicly available.

Conclusion

A responsive, capable, and agile Coast Guard is an indispensable instrument of national security and, in many cases, the first line of defense for America's vast maritime borders. To ensure the Coast Guard is equipped to address the demands of a rapidly evolving operational environment, the Service, with the continued strong support of the Administration and Congress, will maintain momentum for ongoing asset recapitalization programs while embracing decisions based on the principles of strategic management.

The progress the Coast Guard has made in critical recapitalization efforts is tremendous. However, the Service's greatest strength is undoubtedly its people, and Coast Guard operations require a resilient, capable workforce that draws upon the broad range of skills, talents, and experiences found in the American population. Recognizing that operational platforms are hollow without a capable and proficient workforce, the Service will continue emphasis on talent management by implementing its Human Capital Strategy and its Diversity and Inclusion Strategic Plan. Together, modern platforms and a strong, resilient workforce will ensure the Coast Guard is prepared to meet future challenges.

While many challenges still lie ahead, with the continued support of the Administration and Congress, the Coast Guard's future is bright and it will continue to live up to its motto to be Semper Paratus – Always Ready. Thank you for the opportunity to testify before you today and for all you do for the men and women of the Coast Guard. I look forward to answering your questions.

Question#:	1
Topic:	Necessary Information
Hearing:	Implementation of Coast Guard Programs
Primary:	The Honorable David Rouzer
Committee:	TRANSPORTATION (HOUSE)

Question: Due to a recent DC Circuit Court decision, farmers and ranchers across the country will be forced to report low-level manure odors to the US Coast Guard. It's likely that more than 200,000 farmers and ranchers will be now be required to report to the Coast Guard.

Is this information truly necessary for the Coast Guard to do its job in coordinating an emergency response?

Response: On March 23, 2018, the President signed the Fair Agricultural Reporting Method Act (FARM) Act as part of the Consolidated Appropriations Act, 2018, thus eliminating this reporting requirement. In light of that, the Coast Guard will not be receiving this information, and it will not impact its emergency response coordination. If the requirement were still in place, the information would not be necessary for the Coast Guard to do its job in coordinating an emergency response.

Question#:	2
Topic:	Response Center Impact
Hearing:	Implementation of Coast Guard Programs
Primary:	The Honorable David Rouzer
Committee:	TRANSPORTATION (HOUSE)

Question: What impact will an additional 200,000 farms reporting have on the Coast Guard's National Response Center to carry out its work, particularly considering these reports must be submitted by farmers within a 24-hour period?

Response: As explained above, on March 23, 2018, the President signed the Fair Agricultural Reporting Method Act (FARM) Act as part of the Consolidated Appropriations Act, 2018, thus eliminating this reporting requirement. In light of that, the Coast Guard will not be receiving this information, and it will not impact its emergency response coordination. If the requirement were still in place, the CERCLA animal waste reporting requirement would have likely had a negative impact NRC operational capabilities and delayed critical emergency communications. The NRC must focus on time sensitive emergency response efforts that currently benefit the general public and the environment. The NRC annually fields nearly 55,000 phone calls and processes approximately 30,464 incident reports. It is typically staffed by a maximum of six watch standers at peak hours, each of whom undergoes approximately four months of training to prepare them for this position. The current NRC staffing levels, information technology, and infrastructure are not designed to process a sudden or drastic increase in incident report volume. The prospect of receiving additional incident reports related to air releases of hazardous substances from animal waste at farms would have likely negatively impacted NRC operations. Leading up to November 15, 2017, the day on which industry originally thought it would need to start reporting its emissions related to animal waste, the NRC, which typically receives 100-150 calls per a day, handled more than 1,000 calls a day (a 567%-900% increase). This volume created call wait times of two—three hours, leading to delays communicating incidents to first responders.

Coast Guard: Actions Needed to Improve Data Quality and Transparency for Reporting on Mission Performance and Capital Planning: Statement of Nathan Anderson

Why GAO Did This Study

The Coast Guard, a component of DHS, serves as the principal federal agency responsible for maritime safety, security, and environmental stewardship in U.S. ports and waterways. To ensure that the Coast Guard is effectively fulfilling its missions, agency managers must have accurate information and base decisions on sound analyses for effective program management.

This statement discusses Coast Guard actions needed to (1) improve the quality of data used for program management and (2) improve the transparency of its data for reporting on mission performance and planning.

This statement is based on relevant products GAO issued from June 2014 through December 2017 on Coast Guard strategic planning and management issues, as well as related recommendation follow-up conducted through February 2018. GAO reviewed applicable laws, regulations, policies and guidance. GAO also interviewed Coast Guard officials responsible for administering these programs and obtained information on how they used data to inform decisionmaking. GAO interviewed a range of stakeholders, including federal and industry officials.

What GAO Recommends

GAO is not making new recommendations in this statement but has made them to the Coast Guard and DHS in the past on improving its program management through, among other things, better quality and more transparent data. DHS and the Coast Guard agreed with these recommendations and reported actions or plans to address them.

What GAO Found

GAO's prior work recommended multiple actions to improve the Coast Guard's program management by improving the quality of data it uses to manage and report on its mission performance. Specifically, GAO recommended actions such as collecting more complete data and clarifying the data limitations to facilitate more effective program management. For example, in December 2017, GAO found that more information is needed to calculate vessel safety statistics that could enhance the Coast Guard's knowledge about accident, injury, and fatality trends involving commercial fishing vessels. Having more complete information could be useful to carrying out its marine safety mission, and GAO recommended, among other things, that the Coast Guard ensure that data collected during commercial fishing vessel incident investigations is accurately captured. In 2018, the Coast Guard reported taking initial steps to capture more accurate data.

GAO's prior work also identified areas where the Coast Guard could improve the transparency of the data it uses for reporting on its mission performance as well its capital planning purposes. For example, in an October 2017 report on performance goals, GAO found the Coast Guard's Annual Performance Report (APR) has not been released publicly since 2011. Consequently, there has not been full visibility over performance across all of the Coast Guard's missions. Coast Guard officials stated that a decision by Department of Homeland Security (DHS) leadership to limit the number of performance goals shared publicly had deterred the Coast Guard from public release of its APR. GAO recommended that APRs be available on the Coast Guard's website; the Coast Guard plans to publicly release future APRs. In addition, previous GAO reports found that the Coast Guard's annual 5-year capital investment plan, which projects acquisition funding needs for the upcoming 5 years, did not consistently reflect current total cost estimates or the effects of tradeoffs made as part of the annual budget cycle. GAO made recommendations to help the Coast Guard plan for future acquisitions and the difficult trade off decisions it will face given funding constraints. The Coast Guard agreed, but it is unclear when it will complete the 20-year plan.

Chairman Hunter, Ranking Member Garamendi, and Members of the Subcommittee:

I am pleased to be here today to discuss U.S. Coast Guard (Coast Guard) actions that could improve its program management by improving the quality and transparency of the data it uses to manage its mission performance and capital planning. The Coast Guard, within the Department of Homeland Security (DHS), serves as the principal federal agency responsible for maritime safety, security, and environmental stewardship in U.S. ports and waterways. The Coast Guard's 11 primary statutory missions identified in the Homeland Security Act of 2002, as amended, encompass the agency's important roles and responsibilities.[1]

For example, as part of its marine safety mission, the Coast Guard performs mandatory safety inspections, conducts accident investigations, and promotes accident prevention involving vessels at sea, including for commercial fishing vessels.

To ensure that the Coast Guard is effectively fulfilling its missions, agency managers must have accurate and reliable mission performance information to monitor and track the progress the agency is making toward achieving its goals.[2] We have previously reported that evaluation can play a key role in strategic planning and program management, and that findings supported by sound analyses strengthen decision making.[3]

---

[1] The Coast Guard's 11 missions outlined in the Homeland Security Act of 2002, as amended, as well as other mission responsibilities, are identified in table 1 of this statement. See 6 U.S.C. § 468(a).

[2] The Government Performance and Results Act of 1993 (GPRA), as updated and expanded by the GPRA Modernization Act of 2010 (GPRAMA), requires agencies to establish annual performance goals with target levels of performance to measure progress towards those goals. See Pub. L. No. 111-352, 124 Stat. 3866 (2011) (amending Pub. L. No. 103-62, 107 Stat. 285 (1993)). Although GPRA and GPRAMA requirements apply to those goals reported by departments (e.g., DHS), they can also serve as performance management leading practices at other organizational levels, such as for component agencies (e.g., Coast Guard). In addition, GPRA requires executive agencies to prepare an annual performance report (APR) on program performance for the previous fiscal year, including a discussion of why any performance goals were not met and plans to meet those goals in the future.

[3] GAO, *Designing Evaluations 2012 Revision*, GAO-12-208G (Washington, D.C.: January 2012). This report addresses the logic of program evaluation design, presents generally accepted statistical principles, and

Our findings are consistent with Office of Management and Budget (OMB) guidance on assessing an agency's performance, which emphasizes the need for frequent, data-driven reviews, including data analyses, and identifying steps necessary to increase agency performance.[4] However, we have previously raised questions about whether annual performance goals[5] and reported performance information accurately reflect the extent to which the Coast Guard is accomplishing its mission goals. We have also raised concerns about the Coast Guard's management of its capital asset acquisitions given the impact these assets can have on the Coast Guard's ability to conduct some of its missions.

- In 2014, we reported that better information on performance and funding was needed to address Coast Guard acquisition shortfalls.[6]
- In 2016, we reported that the Coast Guard did not provide field units with realistic strategic goals to allocate their limited resources, and testified on Coast Guard actions to improve its asset allocation process.[7]
- In 2017, we issued five reports and testimonies that collectively underscore the importance of collecting good data and addressing its limitations and being transparent about it for the benefit of congressional decisionmakers and the public.[8]

My testimony today describes some of the findings and recommendations from these reports. Specifically, I will discuss Coast Guard actions needed to (1) improve the quality

---

describes different types of evaluations for answering varied questions about program performance, the process of designing evaluation studies, and key issues to consider toward ensuring overall study quality.

[4] GAO, Managing for Results: Agencies Report Positive Effects of Data-Driven Reviews on Performance but Some Should Strengthen Practices, GAO-15-579 (Washington D.C.: July 7, 2015); OMB Circular No. A-11, Preparation, Submission, and Execution of the Budget, pt. 6, Section 270.3 (July 2017).

[5] Throughout this statement we use the term performance goal, which the Office of Management and Budget (OMB) and Government Performance Results Act (GPRA) of 1993 define as comprising a measure, a target, and a time frame. This is the term that most federal agencies use to assess and report performance. See Pub. L. No. 103-62, 107 Stat. 285 (codified as amended at 31 U.S.C. §§ 1115-11125). However, DHS and the Coast Guard use the term performance measure instead of performance goal to distinguish its performance measures from high level mission goals. Since DHS and the Coast Guard use the term performance measure as also comprising a measure, target and a time frame, they have all the elements of what we consider to be a performance goal, and therefore we evaluated them as such.

[6] GAO, Coast Guard Acquisitions: Better Information on Performance and Funding Needed to Address Shortfalls, GAO-14-450 (Washington, D.C.: June 5, 2014).

[7] GAO, *Coast Guard: Actions Needed to Improve Strategic Allocation of Assets and Determine Workforce Requirements,* GAO-16-379 (Washington, D.C.: May 24, 2016). GAO, Coast Guard: Actions Underway to Develop Acquisition Plans that Reflect New Assets and Improve the Asset Allocation Process, GAO-16-633T (Washington, D.C.: June 14, 2016).

[8] GAO, Coast Guard Cutters: Depot Maintenance Is Affecting Operational Availability and Cost Estimates Should Reflect Actual Expenditures, GAO-17-218 (Washington, D.C.: March 2, 2017). GAO, Coast Guard Recapitalization: Matching Needs and Resources Continue to Strain Acquisition Efforts, GAO-17-654T (Washington, D.C.: June 7, 2017). GAO, Coast Guard Acquisitions: Limited Strategic Planning Efforts Pose Risk for Future Acquisitions, GAO-17-747T (Washington, D.C.: July 25, 2017). GAO, Coast Guard: Actions Needed to Enhance Performance Information Transparency and Monitoring, GAO-18-13 (Washington, D.C.: October 27, 2017). GAO, Commercial Fishing Vessels: More Information Needed to Improve Classification Implementation, GAO-18-16 (Washington, D.C.: Dec 14, 2017).

of data used for program management and (2) improve the transparency of data used for reporting on its mission performance and capital planning.

This statement is based on eight products we issued from June 2014 through December 2017 on Coast Guard strategic planning and performance management, and recommendation follow-up activities conducted through February 2018. To perform the work for our previous reports, among other things, we reviewed applicable laws, regulations, policies and guidance for selected performance goals. We interviewed Coast Guard officials responsible for administering these programs and obtained information on their processes for ensuring data reliability, including performing electronic testing for obvious errors in accuracy and completeness. We determined that data were sufficiently reliable for reporting some information, but in other instances, data reliability limitations precluded us from reporting information, which we describe later in this statement. We also conducted site visits and interviewed a range of other stakeholders, including federal officials and industry representatives. Further details on the scope and methodology for the previously issued reports are available within each of the published products. In addition, after the issuance of our reports and through February 2018 we contacted the Coast Guard to obtain updated information and documentation, as appropriate, on the status of recommendations we made. We conducted the work on which this statement is based in accordance with generally accepted government auditing standards. Those standards require that we plan and perform the audit to obtain sufficient, appropriate evidence to provide a reasonable basis for our findings and conclusions based on our audit objectives. We believe the evidence obtained provides a reasonable basis for our findings and conclusions based on our audit objectives.

BACKGROUND

With a staff of over 47,000 members, the Coast Guard operates a multimission fleet of 201 fixed and rotary-wing aircraft and over 1,400 boats and ships.[9] Operational control of surface and air assets is divided into two geographic Areas (Pacific and Atlantic), within which are nine Districts consisting of 37 sectors and the stations within them. The Coast Guard's program oversight, policy development, and personnel administration are carried out at the Coast Guard's headquarters. As shown in table 1, the Coast Guard is responsible for 11 statutory missions identified in the Homeland Security Act of 2002, as amended.10 The Coast Guard manages these missions through six mission programs, also listed in table 1.

[9] In addition to the 47,000 members of the Coast Guard military component (active and reserve), Coast Guard officials reported that about 8,500 civilians support the Coast Guard in both field and staff positions, and these staff are further supplemented by approximately 26,500 volunteer members of the Coast Guard Auxiliary.

[10] 6 U.S.C. § 468(a).

Table 1. Information on the Coast Guard's Mission Programs and 11 Statutory Missions[a]

Mission program	Statutory mission	Description
Maritime security operations	Ports, waterways, and coastal security (response activities)	Ensure the security of the waters subject to the jurisdiction of the United States and the waterways, ports, and intermodal landside connections that comprise the marine transportation system, and protect those who live or work on the water or who use the maritime environment for recreation.
Maritime law enforcement	Migrant interdiction	Stem the flow of undocumented alien migration and human smuggling activities via maritime routes.
	Drug interdiction	Stem the flow of illegal drugs into the United States.
	Living marine resources	Enforce laws governing the conservation, management, and recovery of living marine resources, marine protected species, and national marine sanctuaries and monuments.
	Other law enforcement	Enforce international treaties, including the prevention of illegal fishing in international waters and the dumping of plastics and other marine debris.
Maritime prevention	Ports, waterways, and coastal security (prevention activities)	Ensure the security of the waters subject to the jurisdiction of the United States and the waterways, ports, and intermodal landside connections that comprise the marine transportation system, and protect those who live or work on the water or who use the maritime environment for recreation.
	Marine safety	Enforce laws that prevent death, injury, and property loss in the marine environment.
	Marine environmental protection (prevention activities)	Enforce laws that deter the introduction of invasive species into the maritime environment, stop unauthorized ocean dumping, and prevent oil and chemical spills.
Maritime response	Search and rescue	Search for, and provide aid to, people who are in distress or imminent danger in the maritime environment.
	Marine environmental protection (response activities)	Respond to oil and chemical spills.
Defense operations	Defense readiness	Maintain the training and capability necessary to immediately integrate with Department of Defense forces in both peacetime operations and during times of war.
Marine transportation system management	Aids to navigation	Mitigate the risk to safe navigation by providing and maintaining more than 51,000 buoys, beacons, lights, and other aids to mark channels and denote hazards.
	Ice operations	Establish and maintain tracks for critical waterways, assist and escort vessels beset or stranded in ice, and remove navigational hazards created by ice in navigable waterways.

Source: Coast Guard. | GAO-18-408T.

[a] See 6 U.S.C. § 468(a). The Coast Guard divides responsibilities for 2 of the 11 missions into separate Coast Guard programs. Specifically, the Ports Waterways and Coastal Security mission encompasses the Maritime Security Operations response activities and Maritime Prevention activities. Similarly, the Marine Environmental Protection mission encompasses the Maritime Response program activities and Maritime Prevention activities. The Coast Guard has other mission responsibilities not explicitly delineated in the Homeland Security Act of 2002, as amended. These include products and services for the Intelligence Community; activities and efforts to support U.S. diplomacy and international relations, Bridge Administration, Great Lakes Pilotage, and other Waterways Management functions supplementary to providing Aids to Navigation.

As part of its marine safety mission, for example, the Coast Guard conducts, among other activities, safety inspections and vessel accident investigations, including those involving commercial fishing vessels, which are part of an industry with one of the highest death rates of any industry in the United States. For each of its 11 missions, the Coast Guard has developed goals and targets to assess and communicate agency performance. The Coast Guard's performance assessment process also includes identifying performance gaps and implementing corrective actions to address unmet performance goals. As part of its process, the Coast Guard is to establish targets for the current and subsequent 2 fiscal years, according to Coast Guard officials. Each target is set by the Coast Guard, but according to the Coast Guard's Annual Performance Report (APR), some are derived from external factors. For example, DHS requires that Coast Guard set a 100 percent target for the *percent of people* in *imminent danger saved in the maritime environment*. Further, several of the Coast Guard's assets used to conduct these missions are approaching the end of their intended service lives. As part of its efforts to modernize its assets used to carry out various missions, the Coast Guard has begun acquiring new vessels, such as the National Security Cutter, the Fast Response Cutter as well as other assets. However, concerns surrounding the affordability of this effort remain as the Coast Guard continues to pursue multiple new acquisitions without long-term planning to guide the affordability of its acquisition portfolio. Figure 1 shows the Coast Guard's Fast Response Cutter and National Security Cutter.

Source: U.S. Coast Guard | GAO-18-408T

Figure 1. The Coast Guard's Fast Response Cutter and National Security Cutter.

COAST GUARD ACTIONS THAT COULD IMPROVE DATA QUALITY

We previously reported on actions the Coast Guard could take to ensure that, among other things, it addresses limitations posed by incomplete data, the use of unrealistic asset performance data, and limitations with its performance goal data, for more effective program management.[11]

Examples of data limitations that we have recommended that the Coast Guard take action on are below.

Improve completeness of mission data. In December 2017, we found that several different federal agencies play a role in overseeing and promoting commercial fishing vessel safety, including the Coast Guard.[12]

As part of its marine safety activities, the Coast Guard conducts, among other activities, safety inspections and vessel accident investigations. Commercial fishing has one of the highest death rates of any industry in the United States and vessel disasters are the leading cause of fatalities among fishers, according to the National Institute for Occupational Safety and Health. However, our December 2017 review found that more information is needed to calculate vessel safety statistics that could enhance the Coast Guard's knowledge about accident, injury, and fatality trends involving commercial fishing vessels.

The Coast Guard collects some data on commercial fishing vessels that operate in federal waters—including a vessel's length and construction date—but data on the population of the active U.S. commercial fishing vessel fleet are not complete. Between 2006 and 2015, the Coast Guard investigated 2,101 commercial fishing vessel accidents that were identified as occurring in federal waters.[13] While the number of accidents generally increased over this time period, the number of injuries and fatalities declined over the same 10-year period. However, we could not assess the number of accidents, injuries, and fatalities by fishery— meaning the area in which a certain type of fish (e.g., shrimp, salmon, crab) is caught—because the Coast Guard's data is not complete.[14]

Further, we were unable to calculate the rates of commercial fishing vessel accidents, injuries, and fatalities, because reliable data on certain information needed to do so—including the total number of vessels that are actively fishing and the fishery or region in which the vessel operates—are either not maintained or are not collected by the Coast Guard or other federal agencies. Having this information could be useful to carrying out the Coast Guard's marine safety mission, which includes enforcing laws to prevent death, injury and property loss in the marine environment.

---

[11] GAO-18-16; GAO-18-13; GAO-16-379.

[12] GAO-18-16.

[13] In this statement, a commercial fishing vessel accident refers to instances where the Coast Guard responded after the vessel's crew reported a marine casualty involving damage to the vessel or other property, injury or loss of life, or harm to the environment. Accidents that did not occur in federal waters include, for example, those that occurred in state waters or misclassified observations. The United States' federal waters—also called the Exclusive Economic Zone—begin at the seaward boundary of each of the coastal states (generally 3 nautical miles from shore) and extends to 200 nautical miles from shore. In the instances of Texas, Puerto Rico, and the Gulf side of Florida, federal waters begin at 9 nautical miles from shore. The area within 3 nautical miles is generally referred to as state waters.

[14] A fishery is the combination of one or more fish stocks, and fishing for such fish stocks.

We recommended in our December 2017 report that the Coast Guard ensure that data it collects during commercial fishing vessel incident investigations is accurately captured. We also recommended that the Coast Guard work with stakeholders to form a working group to determine an efficient means to establish a reliable estimate of the population of active commercial fishing vessels. The Coast Guard agreed with both recommendations, and in February 2018 informed us that it is in the process of developing additional data fields to capture more information, such as the fishery in which the commercial fishing is involved, and is engaging stakeholders to establish an appropriate working group. We will continue to monitor these actions.

Use more realistic asset performance data. In our May 2016 report on Coast Guard strategic planning, we found that the Coast Guard did not provide field units with realistic goals for allocating assets, by mission.[15]

We reported that the Coast Guard's strategic allocations of assets were based on unrealistic assumptions about the performance capacity of its assets and did not reflect asset condition and unscheduled maintenance. This was due, in part, to the Coast Guard not including information from its field units on the actual performance of its assets. For example, agency officials noted that one of its classes of cutters was 50 years old and these cutters were hampered by mechanical failures requiring emergency dry dock repairs, which resulted in reduced availability to carry out their missions during the year. In another example, a field unit stated that based on historical use, it planned for 575 hours per vessel for one type of cutter instead of the 825 hours performance capacity.

Because actual asset use has consistently fallen below asset performance capacities, there is not a direct alignment between the Coast Guard's strategic operational goals and its prospects for achieving those goals. As a result, the headquarters' strategic intent, which is based on asset capacity rather than actual performance, did not provide the field with strategic, realistic goals for allocating assets by mission. Agencies should use quality information that is appropriate, current, complete, accurate, accessible, and timely to achieve objectives and address related risks.[16] We recommended that the Coast Guard incorporate field unit input, such as information on assets' actual performance, to inform more realistic asset allocation decisions. The Coast Guard concurred with this recommendation, and in February 2018 informed us that it plans to address this recommendation through changes to two process documents that are under revision, with an expected completion date in March 2018.

---

[15] GAO-16-379.

[16] GAO, Standards for Internal Control in the Federal Government, GAO-14-704G (Washington, D.C.: Sept. 10, 2014).

Improve performance goal data. In our October 2017 review of Coast Guard performance goals, we reported that the Coast Guard and DHS identified limitations with two of the seven selected performance goals we reviewed, including *the five year average number of recreational boating deaths and injuries.*[17] In particular, officials believe that many recreational boating injuries that do not require hospitalization are not reported to the Coast Guard. These officials believe that the amount of underreporting may vary over time due to changes in industry trends, making it difficult to accurately determine actual injury rates and program performance.[18] We determined that the data for this performance goal was not sufficiently reliable for the purposes of our reporting objectives due to these likely limitations. We found that the Coast Guard did not report the possible extent of these limitations with this performance goal in its fiscal year 2016 APR.

For the other performance goal, the Coast Guard and DHS identified limitations with the *number of detected incursions of foreign fishing vessels violating U.S. waters*, which is publicly reported in DHS's APR. DHS's review of this performance goal, reported in August 2015, raised questions about the validity of this goal—that is, whether it provides a useful measure of the Coast Guard's performance. Specifically, the review noted that this performance goal is intended to measure a deterrence effect, but doing so is inherently difficult and may lead to contradictory interpretations of performance. In October 2017, we found that the data for this performance goal was sufficiently reliable for our reporting objective purposes, but questions remain about its validity. Reliable data is not a useful indication of performance unless it is also a valid representation of the goal being addressed.[19] DHS officials reported that they did not include a discussion of the limitations for this performance goal in DHS's fiscal year 2015 APR because the performance goal met the minimum threshold for data reliability despite the goal's limitations. Coast Guard officials reported they were aware of these limitations and were working with DHS and OMB to improve the performance goal and implement corrective actions within 1 to 2 years. We recommended that the Coast Guard assess the extent to which documentation on performance data reliability contains appropriate information on known data reliability limitations and update these documents, as needed, based on the results of the assessment. The Coast Guard concurred and in February 2018, informed us that it had taken initial actions to address our recommendation. However, our preliminary review of these actions indicates that further action will be needed to fully address our recommendation, such as documenting and reporting the limitations of performance data.

---

[17] GAO-18-13.

[18] For example, officials cited the increased use of stand-up paddleboards as an example of changing industry trends.

[19] GAO, *Assessing the Reliability of Computer-Processed Data*, GAO-09-680G (Washington, D.C.: July 1, 2009).

Coast Guard Actions that Could Improve Transparency of Data for Reporting on Its Mission Performance and Capital Planning

Our previous reports have identified areas in which the Coast Guard could improve the transparency of its data used for reporting on its mission performance and planning.

Improve transparency of data on mission performance. In our October 2017 report on performance goals, we found that the Coast Guard's APR has not been released publicly since 2011 due to a previous DHS leadership decision.[20] Consequently, there has not been full visibility over performance across all of the Coast Guard's missions.[21] For example, one of the Coast Guard's missions—defense readiness—has no goals that are publicly reported or shared with Congress, even though measures related to defense readiness are included in the Coast Guard's APR.

Coast Guard officials stated that they could see the benefit of publicly releasing their APR; however, DHS's decision to limit the number of performance goals shared publicly has so far deterred the Coast Guard from pursuing the public release of its APR. DHS officials told us that the department is concerned about conflicting information that a component's APR might present because it is vetted and produced separately from the DHS APR. However, the lack of transparency regarding performance data shared publicly and with Congress can result in an incomplete picture of mission performance and can limit effective oversight of Coast Guard operations. As a result, the public and Congress may be unable to determine the extent to which the Coast Guard is meeting its missions. We recommended that future Coast Guard APRs be available on the Coast Guard's public website. The Coast Guard concurred with this recommendation and in February 2018, the Coast Guard informed us that it had completed its 2017 APR and are determining an appropriate approach for making it publicly available.

---

[20] GAO-18-13.

[21] To measure mission performance, the Coast Guard uses three types of performance goals and measures established by DHS for performance reporting by the Coast Guard and other DHS components: Strategic goals are used to reflect achievement of missions that are publicly reported in the DHS APR. As part of DHS's APR, these goals are subject to GPRA and GPRAMA requirements. Management goals are used to gauge program results and tie to resource requests that are reported to Congress and publicly available through the DHS Congressional Budget Justification, along with the strategic goals. Operational measures are additional DHS component measures not reported by DHS, but used internally by components to inform management of operations and activities. The Coast Guard has at least 150 additional internal performance measures used to inform management of operations and activities based on our analysis of the Coast Guard's Strategic Performance Directive and Operational Performance Assessment Report.

Improve capital planning transparency. In our previously issued work on the Coast Guard's annual 5-year capital investment plan (CIP),[22] we found that the CIP does not consistently reflect current total cost estimates or the effects of tradeoffs that are made as part of the annual budget cycle.[23] We made several recommendations in recent years intended to help the Coast Guard plan for future acquisitions and the difficult tradeoff decisions it will likely face. The Coast Guard generally concurred with these recommendations and is in various stages of implementation. For example, in 2017 we reported that we have made recommendations that DHS and the Coast Guard take several actions to gain an understanding of what the Coast Guard needs to meet its mission within its likely acquisition funding levels.[24] These recommended actions included the Coast Guard: (1) conducting a comprehensive portfolio review across all its acquisitions to develop revised baselines that meet mission needs and reflect realistic funding scenarios and (2) developing a 20-year plan that identifies all necessary recapitalization efforts and any fiscal resources likely necessary to complete these efforts. For example, in 2014 we recommended the Coast Guard develop a 20-year fleet modernization plan that identifies all acquisitions needed to maintain the current level of service and the fiscal resources needed to acquire them. Without these efforts, the Coast Guard will continue, as it has in recent years, to plan its future acquisitions through the annual budgeting process, an approach that has led to delayed and reduced capabilities.

In 2016, the Coast Guard revised its 2005 Mission Needs Statement, which provides a basic foundation for long-term investment planning that is to serve as the basis for evaluating the effectiveness of various fleet mixes, and inform the Coast Guard's CIP. However, the 2016 Mission Needs Statement did not identify specific assets the Coast Guard needs to achieve its missions, nor did it update the annual hours it needs from each asset class to satisfactorily complete its missions.[25] The Coast Guard has stated it is developing a 20-year Long-term Major Acquisition Plan, but it has not stated when the plan will be completed or what will be included in this plan, such as potential trade-offs that could be made across the Coast Guard's portfolio of acquisitions to better meet mission needs within realistic funding levels. A long-term plan with a tradeoff analysis would facilitate a full understanding of the affordability challenges facing the Coast Guard while it builds the Offshore Patrol Cutter. DHS concurred with our 2014 recommendation, but it is unclear when the Coast Guard plans to complete the 20-year plan.

---

[22] Since 2012, the Coast Guard has been legislatively required to submit a CIP annually to certain Congressional committees, alongside its budget proposal, that includes, among other things, projected funding for capital assets in such areas as acquisition, construction, and improvements needed for the upcoming 5 fiscal years. 14 U.S.C. § 2902.

[23] GAO-14-450.

[24] GAO-14-450; GAO-17-654T.

[25] GAO-16-633T.

Chairman Hunter, Ranking Member Garamendi, and members of the sub- committee, this completes my prepared statement. I would be happy to respond to any questions you may have at this time.

The Government Accountability Office, the audit, evaluation, and investigative arm of Congress, exists to support Congress in meeting its constitutional responsibilities and to help improve the performance and accountability of the federal government for the American people. GAO examines the use of public funds; evaluates federal programs and policies; and provides analyses, recommendations, and other assistance to help Congress make informed oversight, policy, and funding decisions. GAO's commitment to good government is reflected in its core values of accountability, integrity, and reliability.

In: Coast Guard Programs
Editor: David J. Simmons
ISBN: 978-1-53615-426-9

Chapter 2

AUTHORIZATION OF COAST GUARD AND MARITIME TRANSPORTATION PROGRAMS*

Subcommittee on Coast Guard and Maritime Transportation

Committee on Transportation and Infrastructure,
U.S. House of Representatives, Washington DC 20515,
March 31, 2017

SUMMARY OF SUBJECT MATTER

To: Members, Subcommittee on Coast Guard and Maritime Transportation
From: Staff, Subcommittee on Coast Guard and Maritime Transpot1ation
Re: Hearing on the "Authorization of Coast Guard and Maritime Transportation Programs"

PURPOSE

The Subcommittee on Coast Guard and Maritime Transportation will meet on Tuesday, April 4, 2017, at 2:00p.m., in 2167 Rayburn House Office Building to examine the United States Coast Guard (Coast Guard or Service) and maritime transpot1ation programs. The Subcommittee will hear from the Coast Guard, the Federal Maritime Commission (FMC), and the Maritime Administration (MARAD).

* This is an edited, reformatted and augmented version of Hearing before the Subcommittee on Coast Guard and Maritime Transportation, of the Committee on Transportation and Infrastructure, House of Representatives, One Hundred Fifteenth Congress, First Session, dated April 4, 2017.

BACKGROUND

Each Congress the Subcommittee on Coast Guard and Maritime Transpotlation moves an authorization bill principally to authorize appropriations for the Coast Guard and also the FMC. The last reauthorization bill (the *Coast Guard Authorization Act of 2016,* P.L. 114-120), provided a two-year authorization of appropriations for fiscal years (FY) 2016 and 2017, with $9.1 billion authorized for each year. This Congress, we will need to move another authorization bill. Such bill may include: provisions suggested by the Coast Guard to improve its operations and programs; studies or reports to be conducted by the Government Accountability Office to review laws and regulations implemented by the Service; legislative changes to maritime laws; land conveyances; and miscellaneous maritime policy items. Historically, the Subcommittee works with the House Armed Services Committee to authorize appropriations for MARAD in the National Defense Authorization bill which is passed each session of Congress. The Subcommittee retains jurisdiction over non-national security aspects of the merchant marine and when appropriate addresses relevant issues in authorization bills.

UNITED STATES COAST GUARD

The Coast Guard was established on January 28, 1915, through the consolidation of the Revenue Cutter Service (established in 1790) and the Lifesaving Service (established in 1848). The Coast Guard later assumed the duties of three other agencies: the Lighthouse Service (established in 1789), the Steamboat Inspection Service (established in 1838), and the Bureau of Navigation (established in 1884).

Under section 2 of title 14, United States Code, the Coast Guard has primary responsibility to enforce or assist in the enforcement of all applicable federa1laws on, under, and over the high seas and waters subject to the jurisdiction of the United States; to ensure safety of life and property at sea; to carry out domestic and international icebreaking activities; and, as one of the five armed forces of the United States, to maintain defense readiness to operate as a specialized service in the Navy upon the declaration of war or when the President directs.

The Coast Guard is directed by a Commandant, who is appointed by the President with the advice and consent of the Senate to a four-year term. On May 30, 2014, President Obama appointed Admiral Paul F. Zukunft as Commandant of the Coast Guard.

The Coast Guard was appropriated $8.4 billion in FY 2015 and $9.2 billion in FY 2016 in discretionary funding. The Administration requested $8.4 billion for FY 2017 for the Service's discretionary programs. Under the FY 2017 Continuing Resolutions (P.L.

114-223 and P.L. 114-254), the Service was appropriated a total of$5.23 billion through April 28, 2017.

The Coast Guard is composed of approximately 40,000 active duty military members, 7,500 reservists, and 8,400 civilian employees. The Coast Guard's personnel operate a diverse fleet of surface and air assets. Vessels under 65 feet in length are classified as "boats" and usually operate near shore and on inland waterways. The Coast Guard operates roughly 1,000 boats, ranging in size from 12 to 64 feet in length. A "cutter" is any Coast Guard vessel65 feet in length or greater. The Coast Guard has 247 cutters, including harbor tugs, buoy tenders, construction tenders, patrol cutters, and three polar icebreakers (only two of which are operational). Additionally, the Coast Guard maintains an inventory of roughly 192 fixed and rotary wing aircraft.

For over two decades, the Coast Guard has been recapitalizing its fleets of offshore cutters, a multi-billion-dollar effort to procure eight National Security Cutters, 25 Offshore Patrol Cutters, and 58 Fast Response Cutters. These cutters will replace ninety cutters and patrol craft of advanced age: 12 high-endurance cutters; 29 medium-endurance cutters; and 49 110-foot patrol boats.

- National Security Cutter (NSC): the Service estimates the total acquisition cost for eight cutters at $5.559 billion, an average of about $695 million per ship. In FY 2016, the Service was appropriated $743 million, including funding for award and production costs associated with construction and delivery of a ninth NSC. The Administration requested $127 million for FY 2017 for NSC Post Delivery Activities. The FY 2017 Continuing Resolutions appropriated a total of $512.1 million for the NSC program.
- Offshore Patrol Cutter (OPC): the Service estimates the total acquisition cost of the 25 ships at $10.523 billion, averaging about $421 million per ship. FY 2016 appropriations included a total of $89 million for the OPC program, with $70.5 million to be used to exercise the option for Detail Design and to commence Phase II of the OPC acquisition. The Administration's FY 2017 request was $100 million for OPC Detail Design and award of Long Lead Time Materials. The Service received $4 million for the OPC in the FY 2017 Continuing Resolutions. The Service awarded the OPC contract to Eastern Shipbuilding Group on September 15, 2016.
- Fast Response Cutter (FRC): the Service estimates the total acquisition cost of the 58 cutters at $3.764 billion, averaging about $65 million per cutter. A total of 38 FRCs have been funded through FY 2016. FY 2016 appropriations included a total of$340 million for the acquisition of six additional cutters. The Administration's FRC request for FY 2017 was $240 million for four FRCs. The FY 2017 Continuing Resolutions provided to the Service a total of $219 million.

- Polar Icebreaker: the Coast Guard has one active heavy icebreaker, Coast Guard Cutter (CGC) POLAR STAR, and one active medium icebreaker, CGC HEALY. The Service estimates it will cost roughly $1 billion for a new heavy polar class icebreaker. In FY 2016, the Service was appropriated $6 million. For FY 2017 the Administration's request was $147.6 million and the Service received $6 million in the FY 2017 Continuing Resolutions. The Service initiated a reprogramming of S30 million from the OPC account in FY 2017 to support six different industry studies to decrease the timeframe for the completion of the first heavy polar icebreaker. The awards of these contracts will be issued in March, 2017.

Coast Guard assets are supported by the Service's shoreside infrastructure. In its annual budget request, Coast Guard shoreside infrastructure includes both major and minor shore projects and survey and design work. Survey and design work includes funding required for planning, environmental and engineering studies, and real property and land acquisitions for future year shoreside projects. For each FY 2015 and 2016, the Service was appropriated $500,000 for survey and design work; the Administration requested that same amount in FY 2017.

Minor shore construction projects usually require less advanced planning than major projects and generally fall under (1) emergency repair projects (with cost estimates of greater than 50 percent of the replacement value), or (2) minor facility improvements (costing more than $1 million) to adapt to changing or increased Coast Guard missions. For each FY 2015 and 2016, the Service was appropriated $5 million tor minor shore projects. The Service requested that same amount in its FY 2017 budget. The FY 2017 Continuing Resolutions did not appropriate funds for minor shore projects.

Major Acquisition Systems Infrastructure (MAS!) includes shore facility infrastructure modifications, upgrades, and new construction associated with homeporting new or modified cutlers, boats, and/or aircraft. It also includes logistic, maintenance, and training support for new or modified assets. For FY 2015, MASI received $16 million in appropriations and in FY 2016, the Service received $52 million for these activities. The Administration requested $28 million for MASI for FY 2017.

The FY 2017 Continuing Resolutions include a combined appropriation of$117.5 million for Coast Guard's Major Shore, Aids to Navigation (ATONS), and Survey and Design projects. The Service's spend plan for this appropriation is incomplete and will remain that way until the FY 2017 appropriations arc finalized. The Service must also address repairs needed for damage caused along the Georgia and Carolina coasts by Hurricane Matthew, and the estimated cost for these 14 projects is $92.3 million.

In addition, the Coast Guard has a sizable backlog of unmet shoreside infrastructure projects. Its prioritized shore backlog includes 30 projects with a cost of $465 million. Its unprioritized shore backlog consists of 68 projects with a preliminary cost estimate of

$1.08 billion. The Coast Guard's total shore infrastructure project backlog cost estimate is $1.55 billion.

FEDERAL MARITIME COMMISSION (FMC)

The FMC was established in 1961 as an independent agency responsible for the regulation of oceanborne transportation in U.S. foreign commerce. The **FMC** administers a limited antitrust exemption for ocean carriers pursuant to agreements filed with the FMC to ensure competition among carriers. The FMC also enforces laws related to cruise vessel financial responsibility to ensure cruise vessel operators have sufficient resources to pay judgments to passengers for personal injury or death or for nonperformance of a voyage.

The FMC is composed of five Commissioners appointed for five-year terms by the President with the advice and consent of the Senate. The Commission is led by a Chairman designated by the President. The Acting Chairman is Michael Khouri who was appointed by President Trump on January 23, 2017.

In each FY, 2015 and 2016, the FMC received $25.7 million in appropriations, with 74 percent covering personnel expenses and 13 percent for rent accounts. The Administration requested $27.5 million for FY 2017. Under the FY 2017 Continuing Resolutions (P.L. 114-223 and P.L. 114-254), the FMC was appropriated a total of$14.7 million through April 28, 2017.

MARITIME ADMINISTRATION (MARAD)

MARAD was established in 1950. The agency administers financial programs to build, promote, and operate the U.S. flag fleet; manages the disposal of federal government-owned vessels; regulates the transfer of U.S. documented vessels to foreign registries; maintains a reserve fleet of federal government-owned vessels essential for national defense; operates the U.S. Merchant Marine Academy; and administers a grant-in-aid program for state operated maritime academies.

MARAD is led by an Administrator appointed by the President with the advice and consent of the Senate. The position is currently vacant. The Acting Maritime Administrator is MARAD Executive Director Joel Szabat.

In FY 2015, MARAD was appropriated $341 million and, in FY 2016, MARAD received $399 million. The Administration requested $403.1 million for FY 2017 to support MARAD programs. Under the FY 2017 Continuing Resolutions (P.L. 114-223 and P.L. 114-254), MARAD was appropriated a total of $223 million through April 28, 2017.

Witness List

Admiral Paul F. Zukunft
Commandant
United States Coast Guard
(Accompanied by: Master Chief Steven W. Cantrell
Master Chief Petty Officer of the Coast Guard, United States Coast Guard)

Mr. Michael A. Khouri
Acting Chairman
Federal Maritime Commission

Mr. Joel Szabat
Executive Director, in Lieu of the Administrator
Maritime Administration

Authorization of Coast Guard and Maritime Transportation Programs

Tuesday, April 4, 2017
House Of Representatives,
Subcommittee On Coast Guard And Maritime
Transportation, Committee On
Transportation And Infrastructure,
Washington, DC.

The subcommittee met, pursuant to call, at 2:38 p.m., in room 2167, Rayburn House Office Building, Hon. Duncan Hunter (Chairman of the subcommittee) presiding.

Mr. HUNTER. The subcommittee will come to order. Welcome, everybody. Good afternoon and welcome to our first hearing in the 115th Congress. We would have had this earlier, but it snowed out here.

Today, we will review Coast Guard and maritime transportation programs. As we all know, the United States Coast Guard is a critical component of our Nation's defense and homeland security. It is an armed service; and of the five armed services, it is unique.

It is the only one with law enforcement abilities. The Coast Guard has moved between different Federal departments over its history, with some departments being a better fit than others for the Service.

I have ongoing concerns with the Coast Guard being within the Department of Homeland Security. On its face, the Coast Guard should fit comfortably within the Department, due to its role in defense and homeland security. However, when it comes to budgetary support, it appears the Department—or more likely it is the Office of Management and Budget—ignores Coast Guard priorities for Department or other administration priorities.

The Coast Guard's budget has been determined to be nondefense discretionary, placing it in competition against all nonmilitary discretionary spending, despite the Coast Guard being a military service at its core. No other military service has experienced such a disadvantage and been denied budget clarity and foresight like the Coast Guard. This, without question, is a big risk to national security and should compel a more serious budget approach.

There are a number of us, including the ranking member and myself, that are members of both this committee and the Armed Services Committee, who understand the requirements of the Service. When the Service is active in a time of conflict, it works as part of the Navy, but every day its missions are critical to our national defense. I will repeat myself. Every day the Coast Guard's missions are critical to our Nation's defense.

We are nowhere near close to the budget numbers being final. And I look forward to working with the Coast Guard to provide the Service with the funding it needs to do its job.

We will also hear from the Federal Maritime Commission, the FMC. The FMC implements the Shipping Act of 1984 and other shipping-related legislation. The FMC administers a limited antitrust exemption for ocean carriers, to ensure fair competition among foreign and U.S. shipping interests.

The contraction of the ocean carrier industry over the last couple of years has many carriers operating within shipping alliances to reduce operating costs. The FMC oversees agreements that form these alliances, to ensure they adhere to the limited antitrust exemption. Recent action by the FMC has U.S. industry concerned that the limited exemption is being misused. The industry was also rocked by the Hanjin bankruptcy, which created turmoil in the supply chain. The subcommittee is interested in how the FMC assesses agreements and works with industry to prevent other supply chain disruptions and maintain fair shipping practices.

MARAD is also with us today. The subcommittee shares jurisdiction over MARAD with the Armed Services Committee, having jurisdiction over the nonnational security aspects of the merchant marine. The subcommittee understands the critical role U.S. mariners have in supporting domestic shipping operations as well as defense operations, including the Maritime Security Program and sealift. The subcommittee looks forward to working with MARAD on these important issues.

And I now yield to my ranking member, Mr. Garamendi; you are recognized.

Mr. GARAMENDI. I thank you very much, Mr. Chairman. I appreciate your interest in beginning the process of writing a new 2-year Coast Guard maritime transportation authorization bill. I look forward to working with you again and with the committee members in the same successful—that we have had in the past.

I would also like to welcome our witnesses, the U.S. Coast Guard, the Federal Maritime Commission, and the Maritime Administration. It is my expectation that this hearing will be but only the first of several exchanges as we work collaboratively to build and shape U.S. maritime policy to meet the many challenges and uncertainties in the world today. In fact, we are in a very, very turbulent and challenging time in the maritime world.

Foremost, I remain extremely concerned about the reported $1.3 billion, 14-percent cut to the Coast Guard budget. The new administration has floated it. If they are serious, it is a pretty good example about knowing the cost of everything, but the value of nothing.

There is little doubt that the administration heard the outcry from Members of Congress. Mr. Chairman, you were leading all of that. However, if the administration thought the release of its fiscal year 2018 skinny budget would somehow allay our concerns, no, it didn't happen. We remain deeply, deeply concerned about any cut to the Coast Guard.

If there is perhaps one thing we could do to respond to the uncertainties that we confront in the maritime domain, it would be to make certain that the Coast Guard is fully funded even at a higher level than last year.

Furthermore, if we are to be sincere in our commitments to protect our maritime borders as well as to ensure the reliability, safety, and security of the U.S. maritime supply chain, we will need to work tirelessly to prevent any shortsighted budget policy becoming a reality. I believe this is probably our first and foremost task ahead of us, but there are many, many others.

For one, the global oversupply of container ships and the bottoming out of shipping rates have created turmoil in the global container shipping market. Bankruptcies, mergers, acquisitions and on and on, and this going on at a dizzying pace. We need to know: how will this transformation affect robust competition, fair pricing for the maritime transportation services and foreign trade? In addition, what collateral effects on U.S. marine terminal operators, escort tugs, and other marine service providers will result from the alliances and all of the shifting that is going on in that sector?

The status and future of the U.S.-flag fleet in the international trade also remains a great concern, especially whether this fleet and its credentialed mariners that are on board will remain capable of providing the reliable, secure sealift capacity for our military. I bring our attention to last week's House Committee on Armed Services hearing with the U.S. Transportation Command and the concerns that they expressed about that.

In closing, Mr. Chairman, we must begin now to rebuild our merchant marine and the U.S.-flag presence in global shipping. The United States has been, remains, or should remain, a maritime nation. Our future, our prosperity, and our security depends upon it.

I am optimistic that the "build America, buy America" mantra of this administration applies equally to the U.S. maritime industry; and, assuming so, I extend my hand in cooperation to help that come to pass.

Thank you, Mr. Chairman.

I would like also to enter into the record a report from my Maritime Advisory Committee that met out in Vallejo, California, at the California State University Maritime Academy. If I could put that in the record for the edification of not only myself but the staff and anybody else that would like to access that.

Mr. HUNTER. Without objection.

[The report entitled "Priorities of the Northern California Maritime Industry" can be found on pages 78–100.]

Mr. GARAMENDI. Many different recommendations that they made, some of which I am sure will find its way into legislation.

With that, I yield back.

Mr. HUNTER. I thank the ranking member.

And out of sorts here, but we are joined by the ranking member of the full committee. And if Mr. DeFazio has anything to say, feel free, have an opening statement.

Mr. DEFAZIO. Thank you, Mr. Chairman. I appreciate you holding this hearing.

And I would echo the concerns of both the chairman and the ranking member as regards the neglect that we have seen of the Coast Guard's needs in the budget proposals. The rumored $1.3 billion cut to build a stupid wall that won't make our country any safer would just be outrageous. You know, the French built the Maginot Line, the Germans went around it.

You know, even if the so-called beautiful wall is built across some very rugged terrain, obviously, they can tunnel under it, they can use catapults and send drugs over it. They can use drones to send drugs in, et cetera. It is not going to solve any national security problem but it would open up the same route the Germans used to invade France to drug smugglers, human smugglers, and potentially terrorists trying to get a tactical nuclear weapon into the United States, which is we would leave our vast coastal areas, 95,000 miles, and 28,000 miles of navigable inland waterways, it would be open season, because the Coast Guard wouldn't be able to provide the coverage we need. They are struggling now with their current budget to provide that coverage.

So if those cuts are a reality, it would be perhaps the most laughable thing, but serious, unfortunately, about the so-called budget proposals and the skinny budget, which doesn't mention the Coast Guard whatsoever.

And I also want to echo the concerns of Ranking Member Garamendi about U.S. flags. He showed tremendous leadership there. I mean, we can't be the greatest nation on earth and a great maritime nation without any maritime under our own control. And alliances are of tremendous concern. I think we should revisit the antitrust immunity that has been granted under law, limited as it is.

Apparently, it may be that the Box Club, you know, was even exceeding, shall we say, the limited antitrust immunity, but as they were all handed subpoenas, much to their surprise, the meeting at the Four Seasons. And hopefully, we can get to the bottom of what they are really up to.

But in terms of denying ports or marine terminal operators, you know, contracts, when you get down to three alliances or two alliances and alliances controlled substantially by foreign interests, I believe it presents national security issues, certainly economic issues, and it merits more attention by this committee.

With that, Mr. Chairman, thanks for the hearing and I yield back the balance of my time.

Mr. HUNTER. I thank the gentleman.

I am going to introduce the witnesses now, starting with the Commandant of the Coast Guard, Admiral Paul Zukunft; Master Chief Steven Cantrell, Master Chief Petty Officer of the Coast Guard; Mr. Michael Khouri, Acting Chairman for the Federal Maritime Commission; and Mr. Joel Szabat, executive director, in lieu of the Administrator, for the Maritime Administration. Admiral, you are recognized.

TESTIMONY OF ADMIRAL PAUL F. ZUKUNFT, COMMANDANT, U.S. COAST GUARD, ACCOMPANIED BY MASTER CHIEF STEVEN W. CANTRELL, MASTER CHIEF PETTY OFFICER OF THE COAST GUARD, U.S. COAST GUARD; HON. MICHAEL A. KHOURI, ACTING CHAIRMAN, FEDERAL MARITIME COMMISSION; AND JOEL SZABAT, EXECUTIVE DIRECTOR, MARITIME ADMINISTRATION

Admiral ZUKUNFT. Good afternoon, Chairman Hunter and Ranking Member Garamendi and distinguished members of the subcommittee. Thank you.

I thank you, Chairman, especially for your unwavering support of the United States Coast Guard, and I ask that my written statement be entered into the record.

The Coast Guard is first and foremost an armed service that advances national security objectives in ways no other armed service can. It begins with our broad authorities. Over 60 bilateral agreements to enforce rule of law on the high seas and into the territorial waters of foreign nations, where transnational criminal organizations thrive, and they directly threaten our Nation's security.

We are witness to illicit trafficking destined for the United States, exploiting vulnerable transit nations in Central America, spawning violent crime and bringing corruption and sowing the seeds of illegal migration, people fleeing their countries of origin to seek safe refuge here in the United States.

In 2016, we removed a record 201 metric tons of cocaine, and we brought 585 smugglers to justice here in the United States, where we have a nearly 100-percent prosecution rate as compared to the less than 10-percent prosecution rate in their countries of origin.

Our greatest challenge in this campaign is one of capacity. And we must maintain our current pace in recapitalizing the Coast Guard fleet while advancing shore-based unmanned aerial systems to enhance our surveillance capacity. Also, in 2016, we awarded a contract to complete the build-out of our fleet of 58 Fast Response Cutters at an affordable price at Bollinger Shipyards, who delivered the most recent 4 ships with zero discrepancies.

We awarded the acquisition of 9 Offshore Patrol Cutters to Eastern Shipbuilding Group, a downpayment for a program of record of 25 of these capable platforms that meet requirements and, again, at an affordable price. And we are cutting steel at Huntington Ingalls Shipyard today on the ninth National Security Cutter. We stood up an integrated program office with the Navy and recently awarded industry studies to commence the build-out of a fleet of three heavy and three medium icebreakers, all meaningful steps to keep our Nation on an accelerated path to deliver the first heavy icebreaker in 2023.

And, again, thank you, Chairman and this committee, for your leadership and vision in helping us establish this collaborative approach.

We also received our fourth consecutive clean financial audit opinion, and have minimized acquisition growth and also eliminated timeline slippages in our major acquisitions. Yet readiness, modernization, and force structure challenges inhibit our ability to fully address the asymmetrical threats in our increasingly volatile world.

Like the other four armed services, the Coast Guard faces the menacing combination of increased mission demands, years of fiscal constraint, and lost purchasing power. The Budget Control Act has eroded our ability to simultaneously execute our daily missions, maintain our contributions to Combatant Commanders, and preserve our readiness for global contingencies.

The Coast Guard is the only armed service that has been funded below the BCA floor in our annualized operations and maintenance appropriation. Our Service Secretary, John

Kelly, understands this issue and fully supports the President's call to rebuild all the armed services.

And I am working to rebuild our long overlooked fleet of 35 inland construction tenders with an average age of 52 years. Now, this is critical in sustaining our inland river system and overall maritime transportation system that generates $4.5 trillion of commerce on an annual basis. This fleet is critical to our economic and national security.

Finally, we need to grow the Coast Guard. We have had 5 consecutive years of funding offsets in our operations and maintenance account, and over the next 5 years we need to restore 1,100 Reserve billets and bring another 5,000 Active Duty members into our Service, while sustaining our more than 8,500 civil servants.

I appreciate the unwavering support of this subcommittee to address our most pressing needs. With the continued support of the administration and Congress, the Coast Guard will remain Semper Paratus—Always Ready.

Thank you for the opportunity to testify today, and I look forward to your questions.

Mr. HUNTER. Thank you, Commandant, and thanks for what you do for obviously the men and women that you represent right now in this room.

And Master Chief Cantrell, same to you. A happy Coastie is an effective Coastie. Thanks for what you do for looking out for the men and women and their families and their well-being. As you know, when you are on the water or in the air as much as you are, it is tough. So thank you.

Master Chief, you are recognized.

Master Chief CANTRELL. Thank you, sir. And good afternoon, Mr. Chairman and distinguished members of the committee.

It is a privilege to appear before you today to represent the nearly 41,000 Active Duty and 7,000 Reserve members of your United States Coast Guard, who stay on the watch every day protecting our Nation. As one of our Nation's armed services as well as a law enforcement agency, they represent the finest force we have ever had in our history.

In this era of record operational demands and uncertain budgets, these dedicated men and women are charged with maintaining operational excellence around the globe in our own Nation's ports and waterways. Performing our missions in a maritime environment is inherently dangerous. Our people operate in the harshest and many times the most unforgiving conditions imaginable, but they are proud to do so every day, with a devotion to duty that is something to marvel at.

Recent discussions surrounding proposed cuts to the Coast Guard budget have highlighted the importance of a strong relationship with our department. I am happy to say Secretary Kelly has shown an indepth appreciation for the hard work performed by Coast Guard men and women. His dedication to ensuring a predictable, supportive Coast Guard budget that reflects commitment to our members and their families is greatly appreciated.

And I want to personally thank all the congressional delegations that have visited Coast Guard units, whether within your constituency or elsewhere. Visiting our Coast Guard men and women and speaking with them face-to-face gives you the ground truth about the impacts your decisions are having on even our most junior personnel. It is also great for our units as members show you the pride they feel every day.

I also want to thank you for your continued support of our re-capitalization efforts, which are making a tremendous impact on national security. Our new platforms are extremely capable and contribute to the efficient execution of our mission and our service to the Nation.

But we mustn't lose sight of the fact that many of our Coast Guard men and women continue to serve in ships and at stations that are older than their parents or in some cases their grand-parents.

But the age and deteriorating state of our assets and infrastructure aren't the only challenges the members of our Service face. We continue to face quality of life concerns. With many of our workforce serving at remote units, far from bases or other military installations and in many cases high-cost coastal areas, access to adequate and affordable housing, medical care, and child care will always be a challenge. It is critically important we find ways to ensure access to these services in these remote areas, especially for the nearly 70 percent of our members who are married or have dependents.

Many times, in addition to paying benefits, the support programs we provide to members and their families determine how long they remain in our Service. I believe we can all attest that we would not be successful in our careers without the love and support of our families. The sacrifices our members and their families make throughout a typical career are tremendous, yet they continue to volunteer to serve and reenlist and not just as individuals but as a family.

Life in today's military presents unique challenges, and our leadership is deeply committed to providing every Coastie and their family with unparalleled quality of life as they serve this great country. I am pleased to have my wife, Janet, here with me today as a representative of our Coast Guard families and as someone who spends so much of her time devoted to meeting Coast Guard families throughout our Service to ensure their voices and concerns are always heard.

We will continue to work on improving these support programs and services to ensure our workforce, both married and single, remain Semper Paratus. We must invest in them.

It is very easy to look at shiny new ships and aircraft with modern capabilities and assume we have achieved some budget success.

However, we cannot forget those soft costs that are so critical to our people and their families.

Once again, we are grateful for your support as we address these challenges. I promise to remain steadfast in ensuring these and other quality of life programs are addressed in our budgets, as our current and future workforce depend on us to be their voice.

Mr. Chairman, members of the committee, on behalf of the men and women of your United States Coast Guard and their families, I thank you for your continued and very public support, and thank you for the opportunity to discuss some of the highlights and challenges Coast Guard women and men face today. Thank you.

Mr. HUNTER. Thanks, Master Chief. I also want to recognize your wife sitting right behind you. Would you like to say a few words? No, just kidding. OK. You can really attest to the quality of life stuff. She said she wants you gone more, though. That is all right, right?

Thanks, Master Chief.

Mr. Khouri, I know you have got to roll after your opening statement. You are recognized.

Mr. KHOURI. Thank you, Chairman Hunter, Ranking Member Garamendi and members of the subcommittee. Good afternoon, and thank you for the opportunity to present testimony.

And first I want to recognize my fellow Commissioners here today, Commissioner William Doyle and Daniel Maffei. Commissioners Dye and Cordero are traveling today.

With your permission, I will summarize the prepared remarks and I request the written testimony, together with a copy of our fiscal year 2016 annual report submitted today, be included in the record.

[The "Federal Maritime Commission 55th Annual Report for Fiscal Year 2016" can be found online at FMC's website at http://www.fmc.gov/assets/1/Page/55Annual Report.pdf.]

Mr. KHOURI. We are now in the 100th year of operation under the Shipping Act. The FMC mission is to foster and facilitate open, free, and competitive transportation market activity, while protecting the shipping public from unlawful, unfair, and deceptive practices, with a minimum of Government intervention and regulatory cost.

Of recent note, on March 6, the Commission unanimously approved a new rule to reduce the complexity, burden, and cost for the 165 ocean carriers and the 6,200 ocean transportation intermediaries who file 53,000 contracts and over 730,000 contract amendments with the agency each year.

I intend for this deregulatory effort to continue as we take up other pending and proposed issues. After consulting with my fellow Commissioners, I have designated the FMC's managing director, Karen Gregory, as the regulatory reform officer. She is now standing up a task force to execute on the regulation review process.

Two thousand sixteen was, indeed, a full and challenging year for the ocean shipping industry. A series of mergers by vessel operators saw the top 20 carriers consolidated into

17 companies. These consolidations further led to a reshuffling of the four carrier alliances serving the U.S. trades.

By mid-2018, with all announced mergers, there will be 10 companies arranged into 3 operating alliances, as they carry 82 percent of containerized cargo across all U.S. trade lanes. Even with these developments, the vessel liner industry is still relatively unconcentrated.

The top three carriers in the U.S. trades have market shares of 12.42 percent, 12.39 percent, and 10.62 percent, respectively. These are far from market-dominant positions and, in fact, we are not close to a concentrated market in accepted economic regulatory terms.

Individual alliance members do not coordinate on marketing or pricing under their operational agreements. The Commission's ongoing monitoring of all alliance activity serves to ensure that the U.S. container trades remain open and competitive. The alliance operations do create vessel utilization efficiencies and lower cost structures that directly benefit U.S. exporters, importers, and our U.S. consumers.

All that being said, concerns about these trends have been expressed by U.S. cargo interests, our exporters and importers. All these new alliances have been filed at the Commission over the last year. We have strengthened all of our economic review processes and require tighter limits on the scope of each agreement's authority.

Seaport efficiency and port congestion received significant attention by the Commission over the last several years. The supply chain innovation team, led by Commissioner Rebecca Dye, has been developing information-sharing protocols that should prove extremely beneficial, and integrating the global supply chain, and provide a boost to the American economy.

Regarding our budget, the FMC is a small agency with a focused mission and a specialized workforce of 127 dedicated employees. Eighty-eight percent of our budget is rent and salaries. Much of the remainder is fixed overhead. I will focus our resources on the core statutory mission of the agency and, when necessary, I will reallocate resources to address mission-critical needs.

In summary, the maritime container supply chain is commercially and operationally complex. As international imports and exports grow, increased cargo volumes will continue to stretch this supply chain. The FMC's role is to ensure that the market for container shipping services remains open, competitive, free from unlawful, unfair, or deceptive practices, but, again, with a minimum of Government intervention and regulatory cost.

Thank you for your attention, and I will be pleased to answer any questions.

Mr. HUNTER. Thank you, Mr. Khouri.

Mr. Szabat, you are now recognized.

Mr. SZABAT. Good afternoon, Chairman Hunter, Ranking Member Garamendi, and distinguished members of the subcommittee. Thank you for inviting me to discuss the

Maritime Administration's programs to foster and develop the maritime industry in the United States. Our military relies on U.S.-flag ships crewed by volunteer American civilian mariners to move our warfighters, equipment, and supplies whenever and wherever they need to go.

Three-quarters of MARAD's budget goes to national security and ensuring that America has a viable commercial merchant marine, one that employs enough qualified merchant mariners through three programs: The Jones Act, the Maritime Security Program, MSP, and Cargo Preference.

The Jones Act ensures a U.S.-flag fleet in domestic trade by requiring American-built, American-owned, and American-crewed vessels to transport passengers and cargo between U.S. ports. MSP and Cargo Preference guarantee a U.S.-flag fleet in international commercial trade. Combined, these 3 programs support 97 large commercial ships in the domestic trade and 81 large U.S.-flag commercial trade ships trading internationally.

There is also a federally owned surge fleet, the first cargo ships out of port in a mobilization or a crisis. MARAD has three-quarters of the ships in this surge fleet. When mobilized, MARAD's 46 ships are crewed by volunteer civilian mariners from the U.S.-flag commercial fleet. As the commercial fleet has shrunk in recent years, it no longer employs enough qualified available American mariners to sustain a full sealift.

MARAD is preparing a range of options to address the mariner shortfall, as Congress required by the House fiscal year 2017 Transportation, HUD appropriations report language. We are closely cooperating in this effort with our military partners and the U.S. Transportation Command and the Military Sealift Command.

MARAD holds another 50 ships in the National Defense Reserve Fleet. Some are held for spare parts, others are being prepared for recycling, and some can be activated if needed. These NDRF vessels include six training ships used by the State maritime academies. MARAD has worked on a common design to replace the old training ships as they age out. We are also investigating other potential alternatives.

The largest and oldest ship, SUNY Maritime's *Empire State*, is due to age out in 2019. Massachusetts Maritime's training vessel is 2 years behind. Texas Maritime also needs a large training vessel. No decision has been made about the Federal Government's role in replacing or overhauling these training ships.

The education component, representing 17 percent of our budget, is primarily for the U.S. Merchant Marine Academy, the Federal flagship of the maritime academies. Over 20 percent of the officers in the U.S. merchant marine and over 70 percent of the merchant mariners who have a military service obligation come from the Academy. In order to graduate, midshipman train for a year on American commercial and sometimes Federal ships. This unique partnership with industry is a core part of the USMMA training, known as Sea Year.

Last June, former Secretary Foxx made a decision to suspend Sea Year as part of an effort to combat sexual assault and sexual harassment at the Academy. Working with MARAD, industry and labor partners have developed new standards to address sexual assault and sexual harassment.

Companies representing 60 percent of our Sea Year commercial capacity have already met the standards and have resumed training midshipmen. By summer, we will roll out best training practices developed with the industry.

Last June, the Academy was put on warning for accreditation. Seven requirements have to be met to remove the warning. The Academy was required to do more to address sexual assault and sexual harassment on campus and at sea. The other five requirements are primarily about restoring authorities that were taken from the Academy's leadership after financial and operational shortcomings were uncovered in 2008 and 2009. The accreditors visited the campus last week. We anticipate that by July, we will know if our progress has been satisfactory.

Finally, MARAD's transportation programs comprise 8 percent of our budget. Since 2009, the DOT's discretionary grant programs have also provided valuable resources. Among these are TIGER and FASTLANE grants, helping shippers expand and improve ports and intermodal infrastructure. We are also working on programs like Deepwater Port Licensing to export oil and liquefied natural gas.

The Delfin LNG project will be the Nation's first deepwater port export facility to export billions of cubic feet of liquefied natural gas to foreign markets.

Our Marine Highways Program is also helping us to maximize our underutilized maritime assets, rivers, lakes and waterways, to transport more cargo and to reduce congestion on our roads, highways and rails.

That is a thumbnail view of MARAD. I ask that my written statement be entered into the record. Thank you for your time, and I am happy to answer your questions.

Mr. HUNTER. Thank you, sir.

We will start with questions. I am going to start with myself. I recognize myself here.

OK. Commandant, here is the opening, I guess my opening that kind of overstates and underscores what is happening in this administration and why you are here today.

Tomorrow, there is a meeting on our other committee, the Armed Services Committee. There is a meeting with all the Service Chiefs except for you. So every Service Chief will be there of the five military services, of which you said in your opening statement you are first and foremost a military service. You are not going to be in that hearing. Why is that? Why are you here and not in the Armed Services hearing tomorrow with your Service Chief peers? Instead, you are here with the FMC and MARAD. It seems incongruent.

Admiral ZUKUNFT. I will answer the question as frankly as possible. Four percent of my budget is funded by defense discretionary. Ninety-six percent of it is nondefense discretionary, which then comes under the oversight of these other committees.

So I typically testify with the other Service Chiefs when it comes to matters of personnel. And the last time we had a Commandant testify with the other Service Chiefs was with the repeal of "don't ask, don't tell." But on budgetary matters, we do not have a seat at that table.

Mr. HUNTER. Well, when it comes to national security, when it comes to the national security plus-up right now, which is not being called that, it is mostly national defense, meaning the other four services, I think which hearings you sit in and which table you sit at plays a role in how you are perceived by the military establishment, by the Department of Defense, and by the American people. So, I mean, I guess this: Would you want to be at that table tomorrow?

Admiral ZUKUNFT. Yes.

Mr. HUNTER. So let's take this to where—you talked about increasing personnel. You said increasing 5,000 personnel. Let me see. This NDAA, this Authorization Act, of which you are not included, only a very small part of it, because you are nondefense discretionary instead of defense discretionary.

Let me see. The bill increases the end strength of the Army. The Army was going to go down to 460,000. It keeps it at 476,000. So basically, it increases 16,000. The Air Force got an Active Duty increase to 321,000, of about 4,000; and the Marine Corps was authorized to bump up of an increase of 3,000. That is all happening now. Like that is happening for the next fiscal year. That is going to be included in this National Defense Authorization Act.

Have you requested to move to expand your numbers?

Admiral ZUKUNFT. When the original Executive order came out, I did send a letter to the chairman of the Joint Chiefs, and provided a list of plus-ups that we would need that meet defense requirements. And part of that is an increase to recapitalize the six patrol boats we have over the northern Arabian Gulf. So we have submitted a request. I have not received a reply.

Mr. HUNTER. You asked for six FRCs from CENTCOM? Is that what you were referring to?

Admiral ZUKUNFT. Yes.

Mr. HUNTER. To be included in OCO funding?

Admiral ZUKUNFT. Right. And any additional plus-ups, any supplemental funding.

Mr. HUNTER. And what have you seen to either of them saying yes or no to those six FRCs?

Admiral ZUKUNFT. I have not received a reply.

Mr. HUNTER. So you don't know yet whether that is going to be included in their OCO request?

Admiral ZUKUNFT. I do not.

Mr. HUNTER. OK. Going back to, there was an article a few days ago where Admiral Michel, the Vice Commandant, said that "there was only one skinny budget"—

this is a quote—"that was delivered to Capitol Hill and when you look at that, that's a sustainment budget for the Coast Guard. That's something we can work with and that's what we're marching forward on."

The other four services are requesting a lot more than you are.

They don't want a sustainment budget. And that goes in line with the President saying that we are going to build up the U.S. military. Yet your wise Commandant is saying that you are fine with a sustainment budget right now, that you are not asking for an increase. I guess my question is, which one is it?

Admiral ZUKUNFT. Well, a sustainment budget still funds us below the Budget Control Act. So I cannot take delight in a budget that continues to fall short of our annualized requirements and operations and maintenance, and also the critical need to continue the pace with which we are recapitalizing, our current level of service, our old ships, and underfunded and undermanned service.

Mr. HUNTER. So you say that Admiral Michel was not speaking on behalf of the Coast Guard, just his view?

Admiral ZUKUNFT. I would say part of that is taken out of context. It was looking at the potential of a 14-percent reduction and to remove that lodestone from around our neck, but it still leaves us stooped over with the other burdens that we have to carry.

Mr. HUNTER. So you want more than a sustainment budget? Admiral ZUKUNFT. I do.

Mr. HUNTER. My last thing is something we have touched on a whole lot. Icebreakers. Let's talk icebreakers. You only have one U.S.-made Jones Act icebreaker in the United States, the *Aiviq*. And I know we have talked offline and online about this. I think Congress is prepared to support trials, longer term trials if Homeland Security is willing to support you by starting this process off to build the requirements and do the operational evaluation and R&D so you can build the requirements so you know exactly what you want in an icebreaker. Where are we in that?

Admiral ZUKUNFT. So we have written letters. The most recent was written on the 21st of February to the owners of the *Aiviq*. We have provided them with criteria of what sea trials would entail. And we would really like to sit down, you know, with the owners at the bargaining table to look at what all of this would entail.

More importantly, what are the costs? Because the cost is of great concern to me. And if we are going to go any further than this, we will need top-line relief to be able to move forward. So this is a funding availability as much as anything else, but we are ready to roll up our sleeves, work with the offeror, and then walk them through.

We have been in this business for going on 70 years of icebreaking. They are new to the business. We know a lot about it, and I think a lot of value can be made by us sitting with the owners with our experience in what this platform would need to be able to do to support our national security objectives.

Mr. HUNTER. What is preventing that, by the way? What is preventing the sit-down with the owners?

Admiral ZUKUNFT. We are waiting. I have written letters. We have received a few emails. But we are reaching out to the owners to have this discussion, which is really where I would like to have this discussion, at the bargaining table. And preferably, we can take some of that burden off your shoulders as well. But we have already committed ourselves in writing to have these discussions.

Mr. HUNTER. Have you gotten any reassurances from Homeland Security that the big Department is willing to find funding for this if it works?

Admiral ZUKUNFT. No, we have not had that level discussion yet. Until, again, we can sit down with the owner and look at—we haven't even talked dollar figures of what all is entailed. So we need to do our homework first before I can present a package to the department.

Mr. HUNTER. Very quickly, Mr. Khouri, if you don't mind, we will submit questions to you for the record. Is there anything else you want to say besides your opening statement?

Mr. KHOURI. I just want to say not only for the record, but commit to every member of the committee that I would be happy to come and meet personally and answer any and all questions that they may have. And I have listened to every one of the opening statements. So, please, we will be available to answer every question. Thank you. I just have a hard stop and must leave to catch a flight.

Mr. HUNTER. I got you. And hopefully, for you and Mr. Szabat——

Mr. DEFAZIO. Is he departing, Mr. Chairman? Mr. HUNTER. He is.

Mr. DEFAZIO. OK. That is unfortunate, because I do have some questions and very serious concerns about their actions regarding these alliances. So I would appreciate——

Mr. HUNTER. I would yield to Mr. DeFazio.

Mr. DEFAZIO. Do you have a couple of minutes? I can have at it fast.

Mr. HUNTER. He has got to leave at 3:30.

Mr. DEFAZIO. And you can give brief answers. Thank you. Thank you, Mr. Chairman.

So now we are going to have these foreign alliances controlling 90 percent of the container market to the Pacific States. You know, aren't you concerned about the potential for antitrust collusion here? Yes or no, maybe?

Mr. KHOURI. There is—we need to understand——

Mr. DEFAZIO. No, I understand the limitations of your authority, but you could have disapproved and had legal action.

So the question is, you let it go through, what are you going to do to mitigate it? You have told some of our operators—this is another question—that the tug people shouldn't worry about their inordinate influence, because you are going to review any jointly negotiated contract between the alliance and U.S. tugboat operators.

The FMC, I am not aware has the authority to do that. Do you have the authority to do that?

Mr. KHOURI. Can I answer the first question first and then second question?

Mr. DEFAZIO. It has got to be brief, so yeah.

Mr. KHOURI. One needs to understand the nature of these alliances. They are purely operational. They do not have any price-setting agreement or discussion in any shape, form, or fashion.

Mr. DEFAZIO. The Box Club who were just all subpoenaed, what were they doing?

Mr. KHOURI. Well, Congressman, with all due respect, you must know more about what the Department of Justice is doing than I do, because I have no idea what Department of Justice——

Mr. DEFAZIO. Well, they have these famously—I mean, there is no one in the industry who thinks these people aren't getting together in the room and colluding over pricing and who is going to get—who is going to control what harbors, what marine facilities, who they are going to—what they are going to do. I mean, it is Pollyannaish to think, oh, these alliances are just to help make the industry more efficient. Twenty years ago that might have been true. It is not true today, and they are foreign-controlled.

Mr. KHOURI. If I could just go through a few numbers very quickly for you without going into weeds and not to be argumentative, Congressman. And you and I have met earlier and I am more than happy to come back to your office and spend as much time with you and staff as may be necessary.

But as we go down, there are 5,000 container vessels in the world. There are 2,600 vessels that operate, are owned by the companies in the alliances. The alliance activity itself is only 887 vessels, meaning all the rest of the vessels are not in alliance activities. There are only 578 vessels for all of the alliances together in the U.S. trades. That means that you have over 2,000 vessels.

Now, one of the core antitrust principles is, is the market contestable and is there entry that can come in, in case prices go up and become too remunerative. Are vessels going to come in and discipline that price. There are, again, 578 alliance vessels serving U.S. trades and there are 2,000 vessels to discipline the U.S. worldwide, there are 4,000 vessels available to come in and discipline the trades.

Mr. DEFAZIO. Can I just for a moment—if you want to keep talking. But can we focus on where these vessels are serving? I believe a large number of the alliances happen to be serving China, the alliance vessels, and Japan.

Doesn't that create some questions? I mean, when Chinese also have—you know, they are going to control who can come in and come out. So we now have a Chinese-controlled alliance, and they are going to want to favor that, you know, as they do famously.

And so, you know, I mean, there is this, oh, all these other people can come in, except, well, the Chinese probably aren't going to load their ships or let them into their harbors. You know, I just think that this is really going too far. I mean, at what point will we think we are too concentrated, when we get to two or one alliances——

Mr. KHOURI. Well, sir——

Mr. DEFAZIO [continuing]. That controls everything?

Anyway, Mr. Khouri, I realize you have got to go. We can perhaps have another meeting to discuss this. But I really think that we need to revisit the act. We need to revisit the assumptions that we are creating efficiency in market forces here. In these modern times where we know that, you know, State-owned enterprises and governments that are acting in a mercantilist way are not really interested in competition, and they are interested in driving down their costs and dominating our markets and putting our people at disadvantage. That is my opinion. You may have a totally different one.

Thank you, Mr. Chairman.

Mr. KHOURI. If I could, just one last comment, Mr. Chairman, and I will answer all of the questions in writing with much more specificity.

But in terms of concentration, it was in my written statement and my oral statement. The transpacific trades, China, all of Asia over to the U.S. west coast. The standard initial measurement that every regulator uses, the Department of Justice, Federal Trade Commission, and the Federal Maritime Commission is what is called the Herfindahl-Hirschman Index, and it measures concentrations in markets.

The transpacific—now, let me—where the standard is, anything from zero to 1,500 is deemed safe harbor, sufficiently unconcentrated that it does not normally warrant scrutiny. From 1,500 to 2,500 is mildly concentrated, and then above.

Our Bureau of Trade Analysis says in 2016, the transpacific HHI was 647. Worldwide, the HHI is 849. This is saying that these markets are still unconcentrated. Now, that is only the beginning of a long series of questions that have to be answered in an antitrust review.

I would be more than happy to sit down with you. But I am not trying to be argumentative. I am just trying to say that there are traditional ways that whomever you may ask, Department of Justice, FDA or—excuse me, FTC or us, these are the principles that if we are to go now into court in front of a Federal judge and say, Your Honor, these are the economic indicators we have, and it is the judge, not us, who issues the injunction.

So I have to have a solid economic argument to go in front of a Federal district judge to convince him or her that we have an economic situation. These are the numbers that we are working with.

Happy to engage in this to your satisfaction, sir.

Mr. DEFAZIO. Well, I have got to say just the marine terminal operators, the tug companies and others who are U.S. residents, U.S. citizens, and operating to the benefit of our country really don't share those views and think it is going to be much more of an antimarket force. So thank you.

Mr. KHOURI. Thank you, Mr. Chairman.

Mr. HUNTER. And it would be maybe strip out the limited antitrust exemption that FMC can grant these consortiums totally, so that they aren't allowed to join together to put pressure on the ports, collude on price, which you say they don't do, et cetera.

But that is it, Mr. Khouri. Thank you.

With that, I would like to recognize Mr. Graves.

Mr. GRAVES OF LOUISIANA. Thank you, Mr. Chairman. I want to thank all of you for being here today.

Admiral, I would love to talk to you a little bit about some of the acquisition efforts underway. Obviously, I think all of us up here have strong concerns that the men and women of the Coast Guard have the vessels and the equipment that they need to properly do their job in an increasingly challenging and demanding work environment that many of our Coasties are subjected to.

One thing I would love to talk to you about a little bit is the FRC. I know some folks just recently went down, including the ranking member of the subcommittee just went down to take a look at that. You mentioned in your testimony in the last four vessels that were delivered, there were no discrepancies. Is that correct?

Admiral ZUKUNFT. That is correct.

Mr. GRAVES OF LOUISIANA. Is that common?

Admiral ZUKUNFT. Rare.

Mr. GRAVES OF LOUISIANA. Cost per copy, how is that looking compared to estimates?

Admiral ZUKUNFT. So we renegotiated phase 2, and this came in at a very affordable range.

Mr. GRAVES OF LOUISIANA. So cost going down, no discrepancies.

Do you see the—and you recently, as you know, awarded Eastern Shipbuilding Group to do the OPCs. Do you see benefits to the Jones Act in regard to the quality of vessels that you are getting?

Meaning because of these companies' efforts to build domestic vessels and just making sure we have a robust defense industrial base, do you see benefits to the Coast Guard, to the Government, and the American public as a result of that?

Admiral ZUKUNFT. Congressman, absolutely. I have been to Bollinger Shipyard as well. And even in this job market today, I wouldn't call it jobs, I would call it skills. And there are still certain skills that are lacking in this country if we ever found an opportunity where we need to increase our shipbuilding capacity, but it is imperative that we continue to retain these skills and, more importantly, continue to build these skills as a maritime nation.

Mr. GRAVES OF LOUISIANA. Thank you. I will move on a little bit, but stay in the shipbuilding realm. The Coast Guard authorization bills, the last few of them have included provisions that require that smaller fishing vessels now comply with the standards of different classification societies. As a result of that, in Louisiana, the second biggest commercial fishing industry in the United States, we have not seen new vessels being built, none, because in many cases they are telling us it is cost-prohibitive.

I have got concerns about two different things there:

One, whether those standards are appropriate for fishing vessels.

Number two, NIOSH, National Institute for Occupational Safety and Health, provides accidents per region for different incidents that happen on fishing vessels. However, it is not broken down by the type of fishing that is being done. And so I am a little bit concerned that this data is not properly conveying the information that we need. For example, you know, is it the "Deadliest Catch" folks that are truly running into the most accidents or problems, or is it "Swamp People" that actually don't really use vessels that are classified. But in any case, I was looking for another show.

But does that make sense? I mean, is it shrimp fishing, is it king crab fishing, and getting a better understanding of which type of vessels or industries are actually having the higher number of accidents where perhaps there should be some intervention or looking to try and reduce accident rates.

Admiral ZUKUNFT. First of all, before there is any intervention, we engage with all stakeholders. We have a fishery advisory committee that we consult with as well. And there are proposed regulations, but, you know, those are being held back right now under an alternative safety compliance plan and what the impact of that would be.

My only concern in all of this—and I have worked in both fleets—is when we have a gulf fleet vessel be reconfigured and now is operating up in the Bering Sea. And so that would be the only case. But, as you mentioned, the casualty rates speak for themselves. It is a safe fishery in the Gulf of Mexico. At the end of the day, you know, we want to not impede their livelihoods.

Mr. GRAVES OF LOUISIANA. Admiral, I would just appreciate if we could get a commitment out of you to work with us on finding the right solution there. I am not sure that some of the Coast Guard provisions that are in law today are necessarily as prescriptive as they should be, based upon the different fishing industries we have in different regions of the country. And I would appreciate it if we could work together with you to try and take a fresh look at that and see if we can find the right solution there.

Admiral ZUKUNFT. Congressman, you have my commitment that we will work with the stakeholders and with the fleet owners as well.

Mr. GRAVES OF LOUISIANA. Thank you. In my last 10 minutes, I just wanted to cover the icebreaker. I know that the chairman noted icebreaking capabilities. You and I have discussed that as well. I think that there is just complete support among the members of this committee about our strong concerns about the lack of icebreaking capability in the United States compared to other Arctic nations.

We have talked at length about the idea of the Coast Guard obviously acquiring additional capabilities, new capabilities. And, again, I think everyone here fully supports that.

The problem that I think most of us see is that there is this interim gap where we simply don't have capabilities. If there were an incident up there, we have already seen Russian aggression up in the Arctic area. Their capabilities, if I remember right, are—what is it?—40 times more than we have in the United States. Extraordinary difference in capabilities.

Chairman Hunter noted the need to develop an interim strategy while we concurrently work on this long-term strategy for acquisition. And I once again would like to reiterate the chairman's comments and support for that, but I would like to hear a commitment from you that we are going to figure out how to work through this and develop an interim strategy.

And I am not going to dictate any type of a path forward on that, but we have got to get interim capabilities, because we simply have insufficient capabilities for this interim period while we concurrently work on long-term acquisition of an icebreaker.

Admiral ZUKUNFT. My reply to that is we are working with one and the only U.S. offeror right now that might be able to bridge this gap. But what if we can't meet that requirement, then where do we turn to? We may have to look at other sources as well if we are, in fact, looking at a lease option.

But, again, we are fully engaged, ready to sit down with the offeror, and put all the pieces on the table of what it would take to make sure that this vessel could meet our requirements in the high latitudes.

Mr. GRAVES OF LOUISIANA. Admiral, I think every time you have come before this committee, this issue has come up. There are strong concerns on both sides of the aisle here about this interim solution. I think all of us are willing to work with you on it.

But I just want to push again that we need to see some substantial progress on finding a path forward for the interim solution here. So thank you.

Admiral ZUKUNFT. Thank you.

Mr. HUNTER. I thank the gentleman.

The ranking member, Mr. Garamendi, is recognized.

Mr. GARAMENDI. Thank you, Mr. Chairman.

Mr. Graves, I almost was in your district at the shipyard.

Mr. GRAVES OF LOUISIANA. No, you were in the district. In between. That is right.

Mr. GARAMENDI. Your boosterism is appreciated and correct. It really was quite an experience.

Mr. GRAVES OF LOUISIANA. The nice area was ours, just to be clear.

Mr. GARAMENDI. Admiral, thank you. Your team did a great job on the codel that we did last weekend, the opportunity to see the Fast Response Cutter and then also over at Panama City, both of them moving along.

I would appreciate a report on the progress in both, particularly the latter, the Offshore Patrol Cutter, where it stands. I notice some of your testimony covered it, but if you could provide some detail on that in the days ahead, it would be much appreciated, or you can give us a quick rundown now.

Admiral ZUKUNFT. I would be pleased to provide that report. And the date to keep in mind is delivery of the first ship in the year 2021.

Mr. GARAMENDI. 2020 for the Offshore Patrol Cutter?

Admiral ZUKUNFT. 2021 for the first Offshore Patrol Cutter.

Mr. GARAMENDI. Very good. We would want to watch that closely and monitor it along and any hiccups or slowdowns. But it was impressive to see what was going on there.

Many, many questions. I want to just take a moment on the budget cuts, proposed, $1.3 billion. Please give me the top three problems that it would create for the Coast Guard.

Admiral ZUKUNFT. Well, I would begin with the acquisition of our ninth National Security Cutter, because that would be removed.

We have already awarded long lead time materials.

Steel is already being cut. We would incur a contract penalty and probably would face a penalty in the amount of what this offset was designed to provide, half a billion dollars. So there is no net savings there.

It would cut our department's only counterterrorism capability.

It has taken us the better part of 6 years, using Seal tactics training procedures, recruiting the best talent, to stand up these teams, and under this proposal we would lose that capability as well. And you don't lose these immediately. These are people, these are billets, and there are dollar figures assigned to them, and they don't leave overnight even if you do a reduction in force.

We enjoy the highest retention rate of any armed service, but our servicemembers look to me as though I have broken faith as their leader. They look to me as their leader, nobody else. That I have broken faith with them. And then what does that do, as we heard from the Master Chief, to the men and the women of the Coast Guard and to their families that their Commandant broke faith? And I will not break faith with our people. But it does have an impact.

Mr. GARAMENDI. I want to dwell on that for just a moment. Chief, you in a conversation indicated that it is important to have the men and women of the Coast Guard parallel, be equal to the benefits, pays, increases in pay, other benefits that the other four branches of the military has. Could you comment on that?

Master Chief CANTRELL. Yes, sir. And we do. We follow suit with DOD on all pay and benefits entitlements. We are eligible to go to their military treatment facilities and all those benefits of being an armed service.

Where I get concerned is we are in a buildup of the military, and we are not included in that discussion. We have a young man or young woman that is considering joining the armed service. If they look to which service, if you would, is being invested in, and they don't see the United States Coast Guard then they choose not to come to our Service. And that is something that as we grow, want to grow over the next few years, that talent is out there.

And we are already in competition with our sister services on getting that right talent, and not just recruiting them but keeping them in our Service, which is why these other housing and medical and child care and those things that some folks just don't think about will make them leave our Service. And we certainly don't want that. But we want to be part of that buildup, sir.

Mr. GARAMENDI. I want to come back to that. And the chairman mentioned the hearing next week with the four other services. Perhaps, Admiral, you need to be invited. So maybe we can do some of that.

Admiral ZUKUNFT. And if I can, Ranking Member, right now our icebreaking program, to recapitalize that is being funded by the Department of Defense. The Coast Guard is at every COCOM. Every geographic Combatant Commander has Coast Guard forces chopped to it as well.

And so yes, we do need to be a part of these discussions if we are talking about building the force for the 21st century.

Mr. GARAMENDI. When Secretary Kelly was the commander of SOUTHCOM, he seemed to think that the Coast Guard was really important, but yet it is his budget that takes $1.3 billion out of your budget. I suspect he—well, let's hope he still thinks that the Coast Guard is important.

Didn't he say that the southwest border is not along the—is 1,500 miles south of the United States border?

Admiral ZUKUNFT. If we talk all borders, our maritime borders begin well south and well beyond and actually east and west as well, thousands of miles.

Mr. GARAMENDI. If you are interdicting drugs off the coast of Colombia, what impact does that have in the Guatemala, Honduras area with regard to immigration from those countries into the United States? Is there an impact?

Admiral ZUKUNFT. So it is insidious the way this works. Colombia, the number one producer of cocaine. I met with President Santos 3 weeks ago. Can they cut back coca production? Can they step up their efforts? It is moved in bulk to Central America. And as soon as it lands in bulk in Central America, corruption goes up, rule of law goes down, and those who can afford a human trafficker will put their young children in a human trafficker to get them into the United States, the number one consumer of cocaine.

So the irony is, it is our consumption that drives the shipment that creates this violent situation in Central America, and yet now people want to migrate to the United States. So this is a—you know, this is much about behavioral health as it is about law enforcement interdiction. It needs to be a campaign, and we need to tackle both of these.

Mr. GARAMENDI. We are going to need to make an argument that a wall on the Mexican border isn't going to stop immigration or the drug issue. It is going to have to be a multiple program, including the Coast Guard in that. We will make that argument a little later.

I do want to go to the maritime—excuse me, not the maritime. But, Mr. Szabat, last week we had a hearing on the Armed Services Committee with General McDew. I believe you are familiar with his testimony in which he spoke to the issue of the pending— not the pending, the real problem that exists today in the lack of mariners and the lack of ships, both the Ready Reserve as well as the MSP. You touched upon that. Is there a solution?

Mr. SZABAT. Congressman Garamendi, thank you for raising that important issue. Short answer is, yes, there is. And I would also say that there is no daylight between us and the U.S. Transportation Command on this issue. In many respects, the Maritime Administration serves as a component command of the USTRANSCOM in providing, you know, 46 of the 61 vessels that they need out of the surge fleet.

Mr. GARAMENDI. We are aging out both the ships, the Ready Reserve ships as well as the sailors on those ships. What do you suggest we do about that?

Mr. SZABAT. Well, Congressman, we have a—as I mention in my testimony, we are, the Maritime Administration is responsible to Congress to come back with a list of options, and we owe it to our new political leadership team as they come on board to vet those options to determine what the recommendations would be. So I cannot speak to our recommendations, but I can speak broadly to what the range of options are.

The policy that we have followed for years in the United States that we rely on the commercial U.S. merchant marine to employ and train enough mariners to serve in both those commercial vessels, but also to generate a surplus that we use on the Federal vessels to meet our sealift requirements.

The challenge that we have today is that the U.S. commercial fleet is no longer large enough to provide both of those needs. So the—one of the two ways forward is, logically, to increase the size of the U.S. commercial fleet. From my testimony, just by doing the numbers, we estimate we are short about 2,000 of these qualified mariners. We need about 40, 45 additional vessels to do that, so anything that would add those 45 vessels.

The two ways that we would do that, one, of course, is the expansion of an MSP or an MSP-like program where we include vessels not just for their military capability, but for their ability to serve in commerce, but also to employ enough mariners so that that pool is large enough to meet our national security needs.

The second way, which would not involve a direct subsidy or direct stipend to the vessels, would be trade policies. Trade policies that would require additional U.S.-flagged vessels, whether they are carrying exports, as the bill that you and Chairman Hunter have proposed, or bilateral agreements as, for example, with China would require a certain number of vessels to sail under the U.S. flag. If you take those approaches, you could also either reduce or eliminate the shortfall of U.S.-flagged vessels. And then—so that is one broad category of increasing the size of the U.S.-flagged fleet.

The second category would be to divorce the need for mariners from the requirement that U.S.-flagged fleet be large enough to employ them. And then the most direct and obvious approach towards that would be to have the United States Navy or the Maritime Administration employ the additional mariners that were needed.

So right now, we rely on the commercial fleet to provide a surplus of 1,300 mariners that we need to draw on within 4 days, within 96 hours of a full sealift activation. The way to divorce that is to say to the Navy or to MARAD: Here are the funds for you to employ 1,300 additional mariners plus the relief crews that you would need going forward in the sealift.

So there were many suboptions and variations within that, but broadly speaking, those are the options before us to address the mariner shortfall and still rely on the U.S.-flagged fleet and U.S. mariners to meet our national security needs.

Mr. GARAMENDI. Mr. Chairman, I am way over my time, so please excuse me. I am just going to wrap it up here very quickly. I would request that General McDew's statement from last week's hearing be entered into our hearing record as it pertains directly to this need for American mariners both for the Ready Reserve Fleet as well as for the MSP. So that would be helpful to understand it.

Also, I draw the attention of the committee to one of the solutions that Mr. Szabat spoke to, and that is, to increase the size of the American merchant marine fleet. And I have some ideas on that, and I have shared that with many of the Members already, and that has to do with cargo and it has to do with trade negotiations. And for example, if we are going to exploit a strategic American resource, such as natural gas and oil, it ought to be on American-built ships with American sailors, thus providing a solution to the overarching problem at hand here.

With that, Mr. Chairman, thank you for the additional time.

[The written statement of General Darren W. McDew, U.S. Air Force and Commander, U.S. Transportation Command, from the March 30, 2017, hearing of the House Armed Services Committee, can be found on pages 101–122.]

Mr. HUNTER. I thank the ranking member.

Let me throw in there too, what General McDew talked about, he said TRANSCOM has never looked at the attrition that would take place with a full sealift callup when you actually have ships sinking. Their numbers do not account for that, and I don't think yours do either.

So they are now modeling what would happen with Korea, what would happen with China, what would happen with Russia, or whoever, if you actually had to call up everybody to deliver gear during wartime, and they are finally using a model that brings actual wartime attrition into that. And I hope you guys are tied in and in sync with that so that your end numbers would come up the same, roughly.

Mr. SZABAT. Mr. Chairman, yes, we are very much in sync with them.

Mr. HUNTER. And by the way, I don't want to take—we are going to come back around to me. I just wanted to throw that out there so that you know what we are looking at.

With that, I would like to recognize Mr. Mast from Florida.

Mr. MAST. Thank you. I appreciate that. Hey, as the only enlisted man on this subcommittee, I just figured it was my duty to harass you for a little while, Master Chief, so I will let you have it here.

I want to get into a little bit into the minutia of your force. I suspect you feel that you feel the finest possible force out there.

Master Chief CANTRELL. Yes, sir, without a doubt.

Mr. MAST. Without a doubt. And as we know, no matter what happens with cuts, what is asked of you is not going to change. People are still going to expect that you step up and meet the mission. You would consider that to be accurate?

Master Chief CANTRELL. Yes, sir. We call it the curse of Semper Paratus. Our folks will find a way, despite budget issues or anything else that is going on. They will find a way to make mission. They know what their job is, and it is something to marvel at. So yes.

Mr. MAST. It is kind of a curse in that respect. You know, many that have never been a part of the force, they might take for granted that the mission gets accomplished because fine folks like yourself always do step up and they accomplish that mission. So like I said, I want to get into a little bit of the minutia of your force. About how many active you have?

Master Chief CANTRELL. Nearly 41,000 active, enlisted and officer.

Mr. MAST. How many cutters?

Master Chief CANTRELL. Oh, gee. Total?

Mr. MAST. Sure.

Master Chief CANTRELL. A couple hundred.

Mr. MAST. A couple hundred. How many boats?

Master Chief CANTRELL. Overall?

Mr. MAST. Yeah.

Master Chief CANTRELL. Several hundred boats.

Mr. MAST. Aircraft?

Master Chief CANTRELL. Again, several hundred aircraft, helicopters.

Mr. MAST. So what does it really look like to that force when you get cut? What does that literally look like to the operational ability of each one of those cutters, those boats? I suspect most of us can agree saltwater is not something that is usually that friendly to the equipment that we go out there and field. What actually gets cut?

What actually happens to your force and your ability to go out there and conduct that cursed mission?

Master Chief CANTRELL. Well, maintenance always suffers when budget cuts happen, and personnel costs, I mean, those things are easy to cut, and they are very expensive. So when we scale back on our personnel and we scale back on the maintenance that they have to provide, those ships and airplanes and small boats get tied up and aren't able to go out and make mission. So we make do with what we have, and it adds extra burden on the folks that are still left around to maintain and do that mission. So they are just—it is added workload for them as they will continue to go out and do what we ask them to do.

Mr. MAST. Very good. What happens with your enlist—you know, we all know, it doesn't matter, Army, Navy, Marines, Air Force, Coast Guard, you got to balance enlistments and retirements. What is going to happen with your force if we get behind the eight ball in balancing that and your ability to go out there and meet the needs of the future?

Master Chief CANTRELL. Right now, we really enjoy a high retention rate in the upper 80s, and it is more than any other of our services. And I believe it is in large part to just people love our Coast Guard mission, they love the way we make effort to take care of them and their families, and they want to stick around.

And I will say personally, I joined the Coast Guard 34 years ago for 4 years, and I am still here, and it is because of the mission and the people and what we do that keep our people coming back. But we have things like blended retirement and things that are on the horizon we have got to pay really close attention to as far as retention as we move out over the next decade. But I think for today, our force is happy to be part of the Coast Guard and will remain happy and will continue to reenlist every time they have the opportunity to.

Mr. MAST. All right. Maybe I am just unique, you know, those of us, especially we come from the coastal communities, we see it in the port infrastructure, the ship traffic, you know, watercraft users, you know, we see your work every single day. I appreciate it, and I appreciate your comments.

I yield back.

Master Chief CANTRELL. Thank you, sir. Thank you for your service.

Mr. HUNTER. Mr. Larsen, you are recognized.

Mr. LARSEN. Thank you, Mr. Chairman. Thanks all for coming.

Oh, there you are. Good. I can see now.

Admiral, can you—I don't want you to forget that there is a maritime border with Canada as well, especially in my district, and I want to emphasize that, and the Coast Guard has never forgotten that. I hopefully helped you not forget that as well.

Recent challenges we are facing there have to do more with Canadian efforts to move oil by pipeline to an expanded pipeline in Vancouver, and then it would be put on ships and come down through Haro Strait and out to Strait of Juan de Fuca, all on the Canadian side. But if something were to occur, result in a spill, obviously, the spill would not be limited to the Canadian side.

So I want to know—I would like to know how aware you particularly are to this challenge we are facing and what Coast Guard is doing about that.

Admiral ZUKUNFT. Thank you, Congressman. And we have regular engagements. We have been having these now for probably close to two decades now with Canada Coast Guard, with Transport Canada, with the RCMP. But to the specific concern that you raise, and especially if it involves tar sands and the like, and when that enters a maritime environment, as you well know, it sinks.

And the oil spill technology is nascent, at best, in terms of the ability to remove a product that is heavier than water, and now it is spreading throughout the environment.

So it is a technical challenge for us. It is a technical challenge for Canada as well. And an oil spill does not respect any borders, and for the very reason that you mention. So it is a concern for me because of the technology it is lacking. We are putting a lot of effort into this oil spill technology, working with oil spill response organizations, whether it is Canada, U.S. it is a concern, though, in your region.

Mr. LARSEN. Are you doing this particularly to the proposed project in Vancouver and British Columbia?

Admiral ZUKUNFT. We need to—yeah, there needs to be an oil spill response plan. There needs to be a contingency plan that goes with that. There needs to be a responsible party. Some of these governance structures, we have the Clean Water Act, Canada does not. Who is going to pay for that removal? So these are a whole list of concerns. And then finally, the ability to remove it, because that is the—really, the biggest challenge that we face right now with these products.

Mr. LARSEN. OK.

Admiral ZUKUNFT. We are not there yet.

Mr. LARSEN. OK. I would like to do some followup with you and your team on that, if I could. I will just note it is April, boating season is starting, so your folks will be even busier in the Pacific Northwest, and I always appreciate the great work that they do both with education of the boating community, the kayaking community as well. But it is coming up, and I look forward to the Coast Guard working with the community up there to maintain the safety of folks out in the water.

Admiral ZUKUNFT. Yeah. Thank you, Congressman. I look forward to working with you as well.

Mr. LARSEN. Great. Thanks.

Is it—sorry, Szabat?

Mr. SZABAT. Yes, sir.

Mr. LARSEN. Great. Thanks. So can you give an update on how many outstanding either letters of commitment or loans, title XI program has right now?

Mr. SZABAT. When you say commitments for title XI, we have at the moment, there are—we have one loan that is in the approval—it has been approved and is in the process of final—of a letter of commitment for us to actually put the money out. And then we have three other interested parties that are applying for title XI loans.

Mr. LARSEN. OK. Then how about in the outstanding side?

Mr. SZABAT. I mean, that universe of four is our universe right now.

Mr. LARSEN. OK. OK. Great. Thanks. And then with regard to small shipyards, there has been $177 million over the last 8 years, 6 or 7 years, allocated for small shipyard grants, including some in my district, including some around the country. None of the shipyards in my area were able to get—meet the standard last—in the last go-round, but I think there was $9 million total, if I am not mistaken. Is that right?

Mr. SZABAT. That is correct, sir.

Mr. LARSEN. OK. And then the proposed budget proposes zero this particular program out. Is that correct? Do I understand that?

Mr. SZABAT. The administration has not submitted a full fiscal year 2018 budget proposal yet, sir.

Mr. LARSEN. All right. Well, if it is anything like the Obama administration, it will be zeroed out, and then we will have to put money back into it.

I think that if I got my numbers right as well on small shipyards, again, $4.9 million in 2016 funding nine projects. But I understand, was there nearly $100 million or so in requests in 2016?

Mr. SZABAT. Congressman, in your typical year, we are going to get somewhere between 10 and 15 times the request, you know, oversubscription.

Mr. LARSEN. In a dollar amount?

Mr. SZABAT. Yeah. So in a dollar amount from, you know, if it is a $10 million program, we will get $100 million or more in requests.

Mr. LARSEN. Yeah.

Mr. SZABAT. And part of that is limited by the fact that the recipients know the program, know that our awards are generally in the $1 million range, so that is what they are asking for. You know, if we had an unlimited amount, obviously they would be asking for more money.

Mr. LARSEN. Sure. And I am not going to assume all those requests are perfect or the best request, but certainly some of that—it is certainly more than what is usually allocated. I guess the point I want to make is that this is a program where there is not just a demand. Demand is infinite, as they say. There is a need, and it does exist and it continues to exist for smaller shipyards around the country and not just where I am from.

So as we move forward to the budget cycle, I certainly will be watching this particular program very closely.

Mr. SZABAT. Congressman, thank you for drawing attention to the program. Your interest is noted, and of course, I will be carrying that back to the Department of Transportation.

Mr. LARSEN. I am sure they are waiting with bated breath to hear. Thank you very much.

Mr. HUNTER. I thank the gentleman.

Mr. Weber for 5 minutes.

Mr. WEBER. Thank you, Mr. Chairman.

Mr. Szabat, is that how you say that?

Mr. SZABAT. Yes, sir.

Mr. WEBER. Pronounce it for me.

Mr. SZABAT. Szabat.

Mr. WEBER. Szabat. OK. You may know I represent the gulf coast of Texas over Louisiana, that other foreign country, and the Texas A&M Maritime Academy is on Galveston Island in my district. We are very proud of that academy and the work going on there, but we are challenged by the limits on their shipboard training capacity. I don't know how familiar you are with the ongoing program there. Do you know that program? You know about it?

Mr. SZABAT. Yes, sir.

Mr. WEBER. OK. Good. We appreciate the support MARAD gives us, provides to the academy. However, our training vessel, *General Rudder*, is just too small. In fact, we are forced to outsource the shipboard training portion of our program to other, what I call the other lesser State maritime academies.

And while we appreciate that support that those academies provide to us, the cost to our academy and our cadets is pretty substantial. Our superintendent, new superintendent, Rear Admiral Mike Rodriguez, tells me that the cost to the academy of outsourcing that shipboard training is more than $2.7 million this year, and that doesn't even include travel cost for the cadets and the faculty or the staff.

So in your view—and I am glad to hear you say you are going to carry a message back to the DOT, but we want you to take it to the guys that are going to make America great again and see how much stroke you have. In your view, what are the next steps toward ensuring that—you are making a note of that. I see that, "take to Donald Trump." What are the next steps in ensuring that each State maritime academy, not just A&M in Texas, but that each maritime academy has shipboard training capacity to fill their missions? What does that take?

Mr. SZABAT. Congressman, you raise a very good and a very complicated issue. This is one that both the State maritime academies and the Maritime Administration have been wrestling with now for some years. I mention in my testimony, past practice has been for the Federal Government almost always to surplus vessels in the National Defense Reserve Fleet to provide vessels to the State maritime academies for training.

Nowadays, if we were to pursue that option, it would be very, very expensive to take an old T5 tanker, for example, and convert her into a vessel that would meet modern safety of life at sea standards, would be about as expensive as purchasing a newly built vessel in a U.S. shipyard. And both of them are—in a tight budget environment, they are both very costly.

Complicating that factor. So right now we have—putting aside Great Lakes Maritime Academy under freshwater, which has issues of its own, you have five State maritime academies essentially sharing four large ships, because as you note, the *General Rudder*——

Mr. WEBER. Say that last line again.

Mr. SZABAT. Five State maritime academies essentially sharing four large ships. As you note, the *General Rudder* is not a large ship.

Mr. WEBER. Yeah.

Mr. SZABAT. And Texas A&M is the fastest growing and will have a need for a large ship going forward.

What complicates this, from my perspective, is the two ships with the largest training capacity, the *Empire State* at SUNY Maritime and the *Kennedy* at Massachusetts Maritime, are the two oldest ships and the two ships that are slated to age out first, so that there is some—there is a time pressure on all of us to identify what the solution is, not just for Texas Maritime, but for all of the State maritime academies going forward.

Mr. WEBER. Well, I appreciate that. We want to know, are you—do you get to have any input with the Office of Management and Budget, OMB, at all?

Mr. SZABAT. It is correct to say, Congressman, that we have an input, but I would go back to Congressman Larsen's comment and say: I am sure they are waiting for my input with bated breath.

Mr. WEBER. Well, now, I gotcha.

Let me jump over to you, Admiral, if I may. My District 14 in Texas, starting in Louisiana and going down not—to Freeport, has five ports, more than any other Member of Congress. Sixty percent of the Nation's jet fuel is produced in our district. It is energy. It is just absolutely used.

The Port of Beaumont, as you may know, moves more military personnel and equipment than any other port in the United States, and that does not include the use of the ship channel in my five ports, by the way. We have the Sabine-Neches Waterway that feeds the Port of Port Arthur and Port of Beaumont. Are you familiar with that area?

Admiral ZUKUNFT. Congressman, I am, and in fact, I went out to Cheniere LNG 2 years ago——

Mr. WEBER. OK.

Admiral ZUKUNFT [continuing]. Before that facility came online.

And we are starting to see an increase in shipping traffic at a brandnew facility as well.

Mr. WEBER. Sure. Well, we would love to see—we have got two Chiefs Reports, you know, on file, if you will, with the Army Corps.

We would love to see both the Sabine-Neches Waterway dredged out and the Port of Freeport as well. So we have got channel improvements on the board.

Do those impact the Coast Guard's ability—when we can't get the Sabine-Neches Waterway dredged to the appropriate depth in Freeport, does that impact your ability to carry out your mission?

You talked about it earlier.

Admiral ZUKUNFT. Precedent does not—we certainly have a responsibility to maintain aids to navigation in these waterways, mark where the good water is, where it isn't. But if it is dredged, then we can anticipate to see increasing shipping traffic.

Mr. WEBER. Absolutely.

Admiral ZUKUNFT. And then it becomes a traffic management concern for the Coast Guard.

Mr. WEBER. OK. Before I yield back, Mr. Chairman, I want to note that we have two LNG plants in our district. We are the 13th largest exporting district in the country out of 435 Members of Congress before our 2 LNG plants come online, which is just almost any day now. And we have a third LNG plant that has applied for permit and on the drawing board.

Mr. Chairman, thank you for the time. I yield back.

Mr. HUNTER. I thank the gentleman.

Master Chief, it is back to me now, and then we will go to Mr. Garamendi when he gets back. I have got—I just want to run this by you, because you are probably the most deeply in tune with the men and women of the Coast Guard, what they are doing, how they live, how their healthcare is, what they eat, how they go to school, so let me ask you this.

Would you say that the men and women of the Coast Guard, when it comes to operational tempo, housing, healthcare, and mission are more aligned, and I am going to use in this instance, the Navy, and I am going to read off some comparisons. Do you think that your men and women of the Coast Guard are more aligned with the Navy or the United States Citizenship and Immigration Services?

Master Chief CANTRELL. Neither. I don't think we are aligned with the Navy or with——

Mr. HUNTER. Does the United States Citizenship and Immigration Services have base housing?

Master Chief CANTRELL. No.

Mr. HUNTER. Do you have base housing?

Master Chief CANTRELL. Yes, sir.

Mr. HUNTER. OK. Do you get BAH?

Master Chief CANTRELL. Yes, sir.

Mr. HUNTER. They don't. The Navy does. The Navy has base housing.

How about are you on TRICARE, or are you on some other kind of civil——

Master Chief CANTRELL. TRICARE.

Mr. HUNTER [continuing]. Healthcare? You are on TRICARE, as is the Navy.

How about CBP, Customs and Border Protection? Do you think that the way that your bases are set up, your operational tempo, your base housing, your healthcare, is it more in line with the military or is it more in line with CBP?

Master Chief CANTRELL. With the military.

Mr. HUNTER. So FEMA. Is it more in line with FEMA or the Department of Defense in the military?

Master Chief CANTRELL. Military.

Mr. HUNTER. We can go through this. Federal Law Enforcement Training Center, United States Immigration and Customs Enforcement, ICE.

TSA. Are you more in line with TSA or the military?

Master Chief CANTRELL. Military.

Mr. HUNTER. OK. I think you see my point here, Commandant, and I will switch to you. You—it seems like we are getting—what we are getting hung up in is would you rather be in the budget pocket of OMB that rules you with an iron fist and treats you as not like a military service to where the plus-up that is happening right now with this administration does not even mention you. All right.

You are talking about budgets, or where do you fit as an operational unit, basically, right? And I think we are getting those two things mixed up. Where would you best fit, if you had your druthers? And I understand you have the first Secretary of Homeland Security that knows, no joke, ran SOUTHCOM, who worked with Coast Guard every day, especially after the Navy left, Coast Guard became the Navy for him in Southern Command, and basically you were the assets for that combatant command in that area.

You have John Kelly right now as your Secretary. There is nobody better, could be than Homeland Security Secretary for the Coast Guard's budget and for just knowing what you all do, and you are still at a sustainment budget. That is with the best Secretary in the world in U.S. history for Homeland Security is in there right now, and you are at a sustainment budget and not mentioned with the other services, right?

Where—explain to me the differences between where you want to sit budgetarily and where you think that you fit in when it comes to your mission, your tempo, your base housing, your men and women, and what you do for this country, national securitywise.

Admiral ZUKUNFT. I would just begin with an annually recurring $2 billion acquisition and construction budget, $2 billion. That is not a lot of money in the Pentagon. We need 5 percent growth in our annualized operation and maintenance, recurring 5 percent growth. We are bringing new assets on board, and it is great that our acquisition budget is being held up. We can't lose sight of the outyear sustainment cost of that.

As a Service Chief—and I sit with all the members of the Joint Chiefs and with the chairman and we help craft a national military strategy, the 4 plus 1 that addresses North Korea, Russia, China, Iran, the Mideast. It doesn't address the Arctic. It does not address the Western Hemisphere.

So I look at where the other services are and then where are they not and then where do our authorities resonate the most, and I have extreme agility. I don't have to go through an ops step, I don't have to go through a tank, I don't have to write a dep board. I just move our ships because I am also the operational commander for the Coast Guard.

I would lose all of that authority if I became a very small component within the Department of Defense. I would much rather be the largest component within the Department of Homeland Security. There will never be a perfect fit. With all of our other statutory authorities, there will never be a—we are a regulatory service. I doubt the Secretary of Defense wants to be dealing with boating safety. He has got higher priorities.

So we always have this regulatory—and a lot of other authorities as well. But our best fit today, Chairman, is with the Department of Homeland Security, but we have got to speak for what our needs are. Part of it is in our culture. We come from a culture of austerity and a culture of offsets, and we need to think of being in a culture of prosperity and growing and not offsetting the Coast Guard, and that is just as much a responsibility of mine as it is for our Service Secretary. And as you mentioned, we have a Service Secretary who gets it, and this is a time for us to come out swinging.

Mr. HUNTER. I mean, you are talking to a U.S. marine here, and they used to—the Marine Corps used to brag about doing more with less. You had 9/11 where operational tempo for everybody, including the Coast Guard, skyrocketed, and they finally realized we can't do as much with less.

They are now asking for the moon, and if they get a little bit less than that, then they are happy with it, and if they have to get what they had last year or just a very small plus-up, they can make do. But your funding requests and the President's funding request in OMB does not match your op tempo and does not match your acquisitions.

If you were in the Department of Defense, I think it would match that. You would be given what you needed to accomplish the missions that are statutory for the U.S. Coast Guard. I think you are kind of stuck somewhere in the middle right now. And I think, if you are going to stay in Homeland Security, then you should get more DOD funding and take on more of a DOD role.

If you go to DOD, you are not going to lose your other core competencies that are regulatory as opposed to warfighting, but if you stay out of DOD, then you will lose that warfighting edge. I think the best analogy that I have read from an article written by the U.S. Naval Institute is you would never send LAPD to Fallujah, but you can take National Guard from Fallujah and put them in L.A. If there is riots, same with Katrina. The military has the ability to come in and do a lot of the natural disasters and humanitarian work that the Coast Guard does. You would team nicely with the Navy as opposed to losing your warfighting capability but keeping those regulatory capabilities in Homeland Security.

I think this is something—I mean, this kind of—people have been talking about this forever, and I—once again, I think you are in the best position you can be in with the Secretary of Homeland Security and this President, and yet you are not seeing—you aren't seeing the rewards from that like other services are. So there has got to be a disconnect, and I think that that disconnect's name is OMB.

Would you like to comment on that?

Admiral ZUKUNFT. Well, how did we get to where we are today?

I will go back over the last 5 years, you know, since BCA came into effect, and we have been funded in our operations and maintenance budget below the BCA floor. So as we put our budget together, we are directed to identify a 5-percent excursion from what is already a minimalist budget, and now take 5 percent off of that as well. And we have done this iteratively for the last 5 years.

I am not aware of my Service Chiefs, as they are doing their budget bill, say, well, now whack everything off the top. In fact, they are getting supplemental funding through OCO.

Mr. HUNTER. In OCO funding.

Admiral ZUKUNFT. And so if we just stop with these offset drills that we do year in and year out, we have squeezed everything out of the Coast Guard. The last time we went through this, I am tying up ships and I am grounding airplanes, because I have already taken everything else out of it. I have shut down pharmacies, I have shut down clinics, I have shut down galleys that affect quality of life at our field units. And every year we go back to say we need 5 percent more, and I can't continue to operate that way.

So one, yeah, we need to change the rule set. This is vital to national security. We can't offset national security. Someone is going to pay the price for it, and I am not willing to pay the price. What has helped us, Chairman, is the work that you have done, and it has been Congress who has restored the Coast Guard, and we will continue to work these relationships.

We have an opportunity with this Service Secretary, who is already engaged at the highest levels, as we look at sustaining the Coast Guard in 2018, but we need to do more than just sustain.

That keeps us on life support. Now we need to grow the Service and not continue to be in the ICU on life support.

Mr. HUNTER. Last question then. How then do you reconcile wanting a plus-up like the other services are now getting and knowing that you are not going to get it because of OMB? How do you reconcile those two things?

Admiral ZUKUNFT. Well, that is predecisional right now, and so we will see how this plays out.

Mr. HUNTER. OMB, first off, they weren't going to leave you sustained. They were going to cut you massively, right. So now they are back to sustainment. But if you were part of the Department of Defense, you wouldn't be seeing that. You would be seeing more money.

So how do you—so how do you say that we fit the best in the organization that keeps cutting us so we can't do our job? I just don't think it—it doesn't make sense to me. So how do you reconcile that?

Admiral ZUKUNFT. Again, we are an instrument of national security. We cannot see these iterative cuts from the top.

Mr. HUNTER. But you are not funded as an instrument of national security.

Admiral ZUKUNFT. And these are the dialogues that we are having right now, you know, with our department, and so we don't find ourselves surprised when we see a passback of this magnitude.

Mr. HUNTER. Is it possible to be included in defense discretionary funding and still be under Homeland Security?

Admiral ZUKUNFT. That is the option that we are exploring right now.

Mr. HUNTER. Say it again?

Admiral ZUKUNFT. We are exploring that option right now.

Mr. HUNTER. What would it take to be considered non—or defense discretionary?

Admiral ZUKUNFT. Defense discretionary. That, I don't know. I mean, it would probably take authorization language to be able to do that.

Mr. HUNTER. That is law?

Admiral ZUKUNFT. Law. To fence off our budget so it does not become subjected to nondefense discretionary offsets.

Mr. HUNTER. What if there were strings attached with that where if you are considered defense discretionary and you get extra money, that that money is put towards those capabilities, which are national security related? Would you have a problem with that?

Admiral ZUKUNFT. I would not. What we do today or every mission set that we do, all 11 of them touch upon national security.

Mr. HUNTER. All right. Thank you.

Mr. Garamendi, you are recognized.

Mr. GARAMENDI. Mr. Chairman, your discussion here is an extremely important one, and it seems to me the solution lies with us. OMB has done its thing, and that is fine. They can suggest, but we are the ones that are actually going to write the appropriation as well as the budgets. And you and I have, for the first time, I think, in a long, long time, you and I also sit on the Seapower and Projection Forces Subcommittee of the Armed Services Committee.

And as members of that committee, I think we have an opportunity here to achieve what you said in your last couple of sentences, and that is to work with the Department of

Defense budget so that some portion of that plus-up, that additional money, is available to meet the needs of the Coast Guard, the national security issues of the Coast Guard, not degrading in any way the national security issues dealing with drug interdiction. But also, you are deeply engaged in what we might call specific defense issues in the Persian Gulf, for example, and other places.

So I think we have an opportunity in that—in our role on the House Armed Services Committee to see that there is a certain level of funding, if it is $54 billion or whatever the additional funding is, that it becomes available to the Coast Guard for a couple of things, and they are on my mind, that the men and women of the Coast Guard have the same benefit package that the men and women of the Department of Defense have so that those needs, and that is housing and all of the rest of it.

Secondly, that there is money for the ongoing development of the icebreakers, not one but three that we need to build in order to meet the needs that we have talked about so many times in the Arctic so that we have set in place in that Department of Defense, National Defense Authorization Act, and in the appropriations and money necessary to carry that out. And the other—and to provide specific money for those things that are specific for the military side of national defense, so that we can parcel out of the budget, whether it is 4 percent or 5 or 10 or 15 percent of your budget and say, OK, here is the OCO funding that you are doing in the Persian Gulf or wherever else that might be.

So I think we have the opportunity, perhaps unique at this particular period of time, because, Mr. Chairman, you and I are on that Seapower and Projection Forces Subcommittee, and we already have a relationship with the Navy with regard to the Coast Guard.

I have asked for some specific information, Commandant, for the scheduling of the money necessary to carry out the Coast Guard program year 1, 2, 3, 4, and so forth, and I want to collapse it into 3 years. You can do it, right, if you had the money?

Admiral ZUKUNFT. Big if. Yes, sir, I am all in.

Mr. GARAMENDI. We understand the difficulties, sir.

I want to move to what is, I guess, my—I don't know, it has become an obsession, how to build ships in the United States, I mean, oceangoing ships.

So, Mr. Szabat, if we could, I want to go through a series of questions with you. You have already spoken to the 78 vessels and probably going down. Has MARAD conducted a formal analysis of the needs and shortfalls it sees in the military sealift for both ships and seafarers?

Mr. SZABAT. Congressman, yes, we have.

Mr. GARAMENDI. Can you broadly describe it, what it is that you—what your analysis is?

Mr. SZABAT. Yes, sir. So just building off of my testimony, and thank you for raising this because from our perspective, this is the single biggest issue that the Maritime Administration and our counterparts in sealift and military side are facing.

So we have identified a shortfall, as I said of about 2,000 mariners, of qualified available mariners in the commercial sector, which equates to, roughly speaking, a shortfall of about 45 vessels sailing under the U.S.-flagged fleet if we continue to rely, as our policy has been and as it makes sense to have a policy to do, to rely on the U.S.-flagged fleet to provide those mariners.

So the first answer to your question, sir, is, yes, we have done the analysis, we have identified the shortfall, and we identified how many vessels would have to be employed, would have to be added to eliminate that shortfall.

However, it does not address a point that Chairman Hunter had made. We are doing this all under the sealift requirements that are established by the Department of Defense, and those requirements, as the chairman mentioned, at the moment leave off two things. One is that they assume that all vessels would be available 100 percent of the time if they are sailing, you know, commercially or activated in wartime, and we know that even in peacetime, no vessel is available 100 percent of the time. There is down time, there is scheduled maintenance, and there is some percentage of breakdowns.

But the other important issue that the chairman identified for going forward for the next war plan is that up until this point, really since the end of World War II, we have always been able to assume no attrition with the merchant marine in our future conflicts, which is a bloodless way of saying that we have been able to assume that we would never have ships torpedoed or just bombed or destroyed by missiles and we would not have fleet merchant marines killed.

Mr. GARAMENDI. OK. I am going to move along here. Mr. SZABAT. Yes, sir.

Mr. GARAMENDI. We are getting near the end of all of this. Can you provide us with both of those scenarios, that is, a peacetime scenario of what our needs are and then a wartime scenario?

Mr. SZABAT. Sir, I cannot provide you with the wartime scenario until we get that from the Department of Defense. I can certainly provide you with the information on the current scenarios that we plan for.

Mr. GARAMENDI. Very good. I would appreciate it if you could do that.

As you know, I have introduced a bill, H.R. 1240, which would expand the U.S.-flagged fleet by requiring a percentage of U.S. crude oil and LNG on American-flagged vessels. We talked to you about this. Would this make a difference?

Mr. SZABAT. Congressman, yes, it would. Mr. GARAMENDI. How and why?

Mr. SZABAT. So the Government Accountability Office indicated that, looking at your proposal, that by around 2025, it would be adding about 100 ships sailing under the U.S. flag for exporting LNG. If I recall correctly, your proposal would be for 30 percent of that to be sailing on U.S. flag, so that would add the equivalent of 30 vessels to the U.S.-flagged fleet, which is a huge cut into our deficit of about 45 vessels that we have today.

Mr. GARAMENDI. Is that for the LNG?

Mr. SZABAT. That is just for the LNG side.

Mr. GARAMENDI. Crude oil, have you done an analysis of what it would mean for crude oil?

Mr. SZABAT. A crude analysis, if you will, Congressman.

Mr. GARAMENDI. Very good. I will take that.

Mr. SZABAT. So currently, we export the equivalent of about one large tanker a day of crude oil, and the challenge here is projecting with the volatility of prices. But if you assume that that stays roughly equal, you are looking at another, you know, 30 or so vessels being employed a year if they were all U.S. flagged. So again, 30 percent of that is another 9 or 10 vessels that would be added to the U.S.-flagged fleet.

Mr. GARAMENDI. So how many, once again, for LNG?

Mr. SZABAT. For LNG, you are looking at 30.

Mr. GARAMENDI. And crude?

Mr. SZABAT. And again, a crude estimate for crude is about another 10, as things stand today.

Mr. GARAMENDI. Another——

Mr. SZABAT. Ten.

Mr. GARAMENDI. Ten.

Mr. SZABAT. Yeah.

Mr. GARAMENDI. All right. So some 40 ships that could be built if we were to pursue such a policy?

Mr. SZABAT. That is what the numbers would say, yes, Congressman.

Mr. GARAMENDI. We heard from Mr. Khouri the issue of bilateral negotiations. I think he mentioned China as an example of a bilateral negotiation. They are set to require that all of the importation of LNG and oil be on Chinese ships. Some of that may be American crude oil and American LNG. I suppose since the President is so intent on bilateral negotiations and he is such a good dealmaker, maybe it could be done.

I think we have covered it here. You did talk about adding sailors with regard—on the military budget, so that is already covered here. I think we have covered those particular issues.

The Ready Reserve capitalization. General McDew said that he would look for ships that are in the worldwide fleet that might be able to replace the Ready Reserve. Would those be American-built ships?

Mr. SZABAT. The ships in the worldwide fleet, Congressman, no, sir. And the challenge with that is just the ships that we have that were built in American shipyards, I think the youngest ship that we have that is sailing currently under a foreign flag is about 34 years old.

So to General McDew's testimony, as I have heard him talk about, is a mix of having some new built in U.S. shipyards, coupled with purchasing foreign-built vessels that would—much younger vessels than the average age of our Ready Reserve force or surge fleet ships today, which are nearly 40 years old.

And I think just to underscore General McDew's point, there has never been a time like there is today in terms of if you want to purchase ships in the international market on the cheap, now is the time to do that. There is a huge surplus on those vessels right now.

Mr. GARAMENDI. We would be purchasing foreign-made ships for a tradition of American-built ships in the Ready Reserve? I think it has been more than the—is it also the law?

Mr. SZABAT. So, Congressman, yes and no. Yes in the sense that we would be building foreign-flagged ships, and yes in the sense that the law currently requires us to have them built in U.S. shipyards. No in the sense that the vast majority of the ships that we have in the Ready Reserve force today were not built in U.S. shipyards. They were brought on board under exceptions or before the law was put in place.

Mr. GARAMENDI. So exception by exception and ship by ship not built in the United States.

Mr. SZABAT. So again, I don't want to speak for General McDew, but from what I have heard from his testimony, he is looking for a cost-effective mix, recognizing that it is hard to build—any proposal to replace, to recapitalize all 61 ships we have in the surge fleet, MSC has 15, we have 46, would be hard to justify for new construction for all 61 of those vessels, especially since all of those vessels, once built, would be going into reserve and would not be going into any sailing service.

So I believe, again, his proposal or his vision is to have that mix of having a good mix of vessels that are built in U.S. shipyards as well as vessels that could be purchased inexpensively overseas to recapitalize the fleet. One way or the other, the fleet does need to be recapitalized.

Mr. GARAMENDI. Well, I for one await his strategy. And I would appreciate your analysis of his strategy and what it means for lost jobs in American shipyards. Apparently, it would not mean for lost sailors on the ship or mariners on the ship, but it would be a problem.

Mr. Chairman, you have already dealt with the issue of the maritime academy ships, which is an ongoing issue that we have, so I won't go to that.

The training vessels. We have heard about the small shipyard grant program. I will say that I was pleased to see how well it operated in the Bollinger yard as well as in the—one of the other yards.

So with that, I think I have covered it. There may be some additional questions that I would like to put for the record at some time. And I thank—Chief, Admiral, Mr. Szabat, thank you so much.

Mr. HUNTER. I thank the gentleman.

I think we are moving forward. I think this is a great time for the Coast Guard. I think that the American people don't know how blessed they are to have leaders like you leading America's Coast Guard. And if we can get you in the right place with the right amount of money, you can do even greater things than you are doing now. So thank you very much for your service and leadership in being here.

Mr. Szabat, thanks for not leaving early. I appreciate it. Maybe and hopefully in the next couple of months we will have more than acting administrators and so forth there. So thanks for being here.

And with that, the subcommittee stands adjourned.

[Whereupon, at 4:28 p.m., the subcommittee was adjourned.]

STATEMENT OF THE HONORABLE JOHN GARAMENDI, SUBCOMMITTEE ON COAST GUARD AND MARITIME TRANSPORTATION, HEARING ON "COAST GUARD MISSION NEEDS AND RESOURCE ALLOCATION," APRIL 4, 2017

Thank you, Mr. Chairman. I appreciate your interest in beginning the process of writing a new two-year Coast Guard and Maritime Transportation authorization bill. I very much look forward to working with you again in that endeavor.

I also would like to welcome our witnesses from the United States Coast Guard, the Maritime Administration, and the Federal Maritime Commission, and thank them for their participation this afternoon.

It is my expectation that this hearing will be but only the first of several exchanges as we work collaboratively to build and shape U.S. maritime policy to meet the many challenges and uncertainties we face today.

The fact of the matter is we are experiencing very, very turbulent and challenging times in the maritime world.

Foremost, I remain extremely concerned about the reported cuts of $1.3 billion- or fourteen percent- to the Coast Guard that the new administration has floated. If ever there was an example of someone "knowing the cost of everything, but the value of nothing," this is it!

There is little doubt that the administration heard the outcry from members of Congress. However, if the administration thought the release of its Fiscal Year 2018 "skinny' budget would allay concerns voiced by Chairman Hunter, me and other members, they were sorely mistaken as they failed to detail any funding for the Coast Guard.

If there is perhaps one thing we could do to respond to the present uncertainties we confront in the maritime domain, it would be to provide certainty in funding for the United States Coast Guard.

Furthermore, if we are to be sincere in our commitment to protect our maritime borders, and as well, to ensure the reliability, safety and security of the U.S. maritime supply chain, we will need to work tirelessly to prevent such a shortsighted budget policy from becoming a reality.

Beyond Coast Guard funding, several different issues are roiling the waters and deserve our attention.

For one, the global oversupply of container ships and the bottoming out of shipping rates have created turmoil in the global container shipping market. Resultant bankruptcies, mergers, acquisitions, and the recent formation of three new vessel-sharing alliances continue this transformation at a dizzying pace.

We need to know: how will this transformation affect robust competition and fair pricing for marine transportation services in the U.S. foreign trade? In addition, what collateral effects on U.S. marine terminal operators, escort tugs and other marine service providers will result from these alliances?

The status and future of the U.S. flag fleet in the international trade also remains a great concern, especially whether this fleet and the credentialed mariners onboard - will remain capable of providing reliable and secure sealift transportation for our military.

In closing, Mr. Chairman, we must begin now to rebuild our merchant marine and the U.S. flag presence in global shipping. The United States has been, and remains, a maritime nation. Our future prosperity and security remains tethered to the sea.

I remain optimistic that the "Build America, Buy America" mantra of this administration applies equally to the U.S. maritime industry. Assuming so, I extend my hand in cooperation to help shape that new future and to reclaim our maritime heritage.

Thank you.

Commandant
United States Coast Guard
2703 Martin Luther King Jr Ave SE
Washington, DC 20593-7000

TESTIMONY OF ADMIRAL PAUL F. ZUKUNFT, COMMANDANT, U.S. COAST GUARD ON AUTHORIZATION OF COAST GUARD AND MARITIME TRANSPORTATION PROGRAMS, BEFORE THE HOUSE COMMITTEE ON TRANSPORTATION AND INFRASTRUCTURE, SUBCOMMITTEE ON COAST GUARD AND MARITIME TRANSPORTATION, APRIL 4, 2017

Good afternoon Mr. Chairman and distinguished members of the Subcommittee. I appreciate the opportunity to testify today and thank you for your enduring support of the United States Coast Guard.

As the world's premier, multi-mission, maritime service, the Coast Guard offers a unique and enduring value to the Nation. The only branch of the U.S. Armed Forces within the Department of Homeland Security (DHS), a federal law enforcement agency, a regulatory body, a first responder, and a member of the U.S. Intelligence Community the Coast Guard is uniquely positioned to help secure the border, combat transnational criminal organizations (TCO), and safeguard America's economic prosperity.

Indeed, the Coast Guard's combination of broad authorities and complementary capabilities squarely align with the Administration's priorities, and I am proud of the return on investment your Coast Guard delivers on an annual basis.

I appreciate the unwavering support of this Subcommittee to address our most pressing needs. I will continue working with Secretary Kelly, the Administration, and this Congress to preserve momentum for our existing acquisition programs and employ risk-based decisions to balance readiness, modernization, and force structure with the evolving demands of the 21st century.

Appropriately positioned in DHS, the Coast Guard is a military Service and a branch of the Armed Forces of the United States at all times.[1] We are also an important part of the modem Joint Force,[2] and a force multiplier for the Department of Defense (DoD). I am proud of our enduring defense contributions to Combatant Commanders around the globe.

In addition to the six cutters operating as part of Patrol Forces Southwest Asia (PATFORSWA) since 2003, other defense operations include:

- Port Security Units (PSUs) support Combatant Commanders with 24-hour protection of vessels, waterways, and port facilities. These specialized teams

[1] 14USC§ 1.

[2] In addition to the Coast Guard's status as an Armed Force (10 U.S.C. § 101), see also Memorandum of Agreement Between the Department of Defense and the Department of Homeland Security on the Use of Coast Guard Capabilities and Resources in Support of the National Military Strategy, 02 May 2008, as amended 18 May 2010.

have deployed almost continuously to strategic ports in Kuwait and in Guantanamo Bay, Cuba, since 2002.

- Deployable Specialized Forces Advanced Interdiction Teams support U. S. Central Command (CENTCOM) vessel board, search, and seizure operations.
- Aircrews perform rotary-wing air intercept operations in support of the North American Aerospace Defense Command (NORAD). Specially trained aviators intercept aircraft that enter restricted airspace in the National Capital Region and during National Security Special Events around the country.
- Assets and personnel deploy worldwide in support of defense operations and fully participate in major international exercises. As the Coast Guard is similar in size, composition, and missions to most of the world's navies, we are a frequent engagement partner of choice to support Combatant Commander goals.

Like the other military Services, the Coast Guard supports all efforts to rebuild the Armed Forces.

Secretary Kelly leads the Department's efforts to secure our borders, and the Administration's strategy "to deploy all lawful means to secure the Nation's southern border..."[3] relies on the Coast Guard supporting a comprehensive security strategy. The Coast Guard protects the maritime border not just here at home, but also off the coast of South and Central America. As Secretary Kelly has stated, "...the defense of the southwest border really starts about 1,500 miles south ..."[4]

We continue to face a significant threat from TCOs, and the Coast Guard is positioned to attack these criminal networks where they are most vulnerable, at sea. We leverage over 40 international maritime law enforcement bilateral agreements to enable partner nation interdictions and prosecutions, and employ a robust interdiction package to seize multi-ton loads of drugs at sea before they can be broken down into small quantities ashore.

In close collaboration with partner Nations and agencies, the Coast Guard works to engage threats as far from U.S. shores as possible. In 2016, Coast Guard and partner agencies interdicted more cocaine at sea than was removed at the land border and across the entire nation by all federal, state and local law enforcement agencies combined. A service record 201.3 metric tons of cocaine (7.1% of estimated flow)[5] was removed from the western transit zone, 585 smugglers were detained, and 156 cases were referred for prosecution.

---

[3] Executive Order on Border Security and Immigration Enforcement Improvements, 25 January 2017.

[4] Secretary Kelly Hearing Testimony, "Ending the Crisis: America's Borders and the Path to Security" before the House Homeland Security Full Committee and Subcommittee on Border and Maritime Security Joint Hearing on America's Borders, Panel 1, 07 February 2017.

[5] [US Department of Homeland Security, Office of Inspector General, *Review of U.S. Coast Guard's Fiscal Year 2016 Drug Control Performance Summary.* OIG Report, OIG-17-33, February 1, 2017.]

Coast Guard readiness relies on the ability to simultaneously execute our full suite of missions and sustain support to Combatant Commanders, while also being ready to respond to contingencies. Your Coast Guard prides itself on being *Semper Paratus* - Always Ready, and predictable and sufficient funding is necessary to maintain this readiness in the future. Prudence also demands we continue investing in a modernized Coast Guard. Indeed, recapitalization remains my highest priority, and today's activities will shape our Coast Guard and impact national security for decades. Your support has helped us make tremendous progress, and it is critical we build upon our successes to field assets that meet cost, performance, and schedule milestones. I am encouraged by our progress to date.

In 2016, we awarded a contract to complete build out of our fleet of 58 Fast Response Cutters- at an affordable price - and the last four ships (numbers 19 through 22) were delivered by Bollinger Shipyards with zero discrepancies. In September, we achieved a monumental goal with the award of a contract for Detail Design and Construction of the Offshore Patrol Cutter (OPC). These cutters will eventually comprise 70 percent of Coast Guard surface presence in the offshore zone. OPCs will provide the tools to more effectively enforce federal laws, secure our maritime borders by interdicting threats before they arrive on our shores, disrupt TCOs, and respond to 21st century threats. With the continued support of the Administration and Congress, we anticipate ordering long lead time material for the first OPC later this year, and plan for its delivery in 2021.

We also generated momentum to build new polar icebreakers. In July of last year I sat before this subcommittee and made a commitment to partner with the Navy to establish an Integrated Program Office to acquire new heavy icebreakers. This approach leverages the expertise of both organizations and is delivering results. The recent award of multiple Industry Studies contracts a concept the Navy has utilized in previous shipbuilding acquisitions to drive affordability and reduce schedule and technical risk is an example of the positive results of this partnership. We will continue refining the system specification and prepare to release a request for proposal for Detail Design and Construction in FY 2018.

We are also making progress with unmanned aerial systems. A recent small Unmanned Aerial System (sUAS) proof of concept aboard a National Security Cutter (NSC) validated this capability and will enhance the effectiveness of these cutters. In its inaugural month underway, STRATTON's sUAS flew 191 flight hours, providing real-time surveillance and detection imagery for the cutter, and assisting the embarked helicopter and law enforcement teams with the interdiction or disruption of four go-fast vessels carrying more than 5,000 pounds of contraband. In addition, we are exploring options to build a land-based UAS program that will improve domain awareness and increase the cued intelligence our surface assets rely upon to close illicit pathways in the maritime transit zone. While long-term requirements are being finalized, I can fully employ a squadron of six platforms outfitted with marine-capable sensors now and am moving out to field this much-needed capability.

In addition to the focus on recapitalizing our surface and aviation fleets, we are also mindful of the condition of our shore infrastructure. Investments in shore infrastructure are also critical to modernizing the Coast Guard and equipping our workforce with the facilities they require to meet mission.

America's economic prosperity is reliant on the safe, secure, and efficient flow of cargo through the Maritime Transportation System (MTS), which sees $4.5 trillion of economic activity annually. The Nation's maritime industry and the MTS face many challenges, including growing demands, a global industry-driven need to reduce shipping's environmental footprint, and the ever-increasing complexity of systems and technology.

Coast Guard marine safety programs employ our unique capabilities to ensure a safe, secure, and environmentally sound MTS. We do this by developing risk-based standards, training and employing a specialized workforce, and conducting investigations into accidents and violations of laws so standards can be improved. We are mindful of the need to facilitate commerce, not impede it, and remain committed to our prevention missions.

While readiness and modernization investments will improve current mission performance, the right force is central to success. I am incredibly proud of our 88,000 active duty, reserve, civil service, and auxiliary members. I am working aggressively to validate a transparent and repeatable model to identify the appropriate force structure required for the Coast Guard to simultaneously respond to global, national, and regional events.

Funding 21st century Coast Guard platforms and people is a smart investment, even in this challenging fiscal environment. Modern assets bring exceptional capability, but our greatest strength will always be our people. Coast Guard operations require a capable, proficient, and resilient workforce that draws upon the broad range of skills, talents, and experiences found in the American population. Together, modern platforms and a strong, resilient workforce will maximize the Coast Guard's capacity to meet future challenges.

History has proven that a responsive, capable, and agile Coast Guard is an indispensable instrument of national security. With the continued support of the Administration and Congress, the Coast Guard will continue to live up to our motto. We will be *Semper Paratus- Always Ready.* Thank you for the opportunity to testify before you today and for all you do for the men and women of the Coast Guard. I look forward to your questions.

Question#:	1
Topic:	Marine Inspection Positions
Hearing:	Authorization of Coast Guard and Maritime Transportation Programs
Primary:	The Honorable Duncan D. Hunter
Committee:	TRANSPORTATION (HOUSE)

Question: The Coast Guard provided 500 new billets and increased funding for training between 2008 and 2012 to enhance the Coast Guard's capabilities to execute its statutory marine safety mission. Industry is concerned that the marine safety and vessel inspection function is degrading, with billets moving out through transfers or being left unfilled.

How many marine inspection positions docs the Coast Guard anticipate for fiscal 2018? How many are military billets and how many arc civilian billets?

Response: There are 670 marine inspections billets in FY18. Of those, 533 are military officers (including warrant officers) and 137 arc civilian marine inspectors. Approximately 37% (250) of the total marine inspector billets are apprentice marine inspectors, and 63% (420) are journeyman and advanced marine inspectors.

Question: Can the Coast Guard provide, for the fiscal years 2013 to 2018, the total number of marine inspection positions, broken down between military and civilian workers?

Response: Between fiscal years 2013 and 2018, the total number of marine inspectors averaged 673 billets. The annual break-down of billets is included below.

Fiscal Year	**Military**	**Civilian**	**Total**
FY2013	536	135	671
FY2014	539	140	679
FY2015	540	136	676
	539	136	675
		137	669
			670

Question#:	2
Topic:	Billet Realocation
Hearing:	Authorization of Coast Guard and Maritime Transportation Programs
Primary:	The Honorable Duncan D. Hunter
Committee:	TRANSPORTATION (HOUSE)

Question: Can the Coast Guard provide a Sector-by-Sector Prevention billct reallocation under the Reprogramming and Optimization of Sector Enterprise (ROSE) initiative? Specifically break out marine inspector and marine investigator billets by sector. Is ROSE complete or are more Prevention billets moving in the future?

Response: Thank you for your interest in the ROSE initiative. The reallocation of billets across the Coast Guard's shore forces field units was a very complex endeavor, involving hundreds of billets across multiple career specialties. At your convenience, my shore forces staff can provide you a brief on the requested details.

Question#:	3
Topic:	Additional billets
Hearing:	Authorization of Coast Guard and Maritime Transportation Programs
Primary:	The Honorable Duncan D. Hunter
Committee:	TRANSPORTATION (HOUSE)

Question: With the addition of nearly 6,000 towing vessels to the U.S. inspected fleet, is the Coast Guard seeking additional billets, both military and civilian. needed to carry out its marine safety mission?

Response: The addition of nearly 6,000 towing vessels to the domestic inspected fleet is an example of the growing capacity on our nation's waterways. The Coast Guard is examining a variety of means by which to meet the challenges posed by this growing capacity. While we evaluate any potential personnel needs, the Coast Guard will continue to adapt our standards and compliance processes, enhance our technical competency, and increase the productivity of our workforce to keep pace with advancements in the maritime industry.

Question#:	4
Topic:	International Maritime Organization's Convention of Standards
Hearing:	Authorization of Coast Guard and Maritime Transportation Programs
Primary:	The Honorable Duncan D. Hunter
Committee:	TRANSPORTATION (HOUSE)

Question: U.S. mariners were required to be up-to-date with their endorsements to comply with the International Maritime Organization's Convention of Standards of Training, Certification and Watchkeeping for Seafarers. Full implementation was required by December 31, 2016. However, the Coast Guard delayed implementation.

Why did Coast Guard delay implementation?

Response: The Coast Guard did not delay implementation.

Question: What impact, if any, does the delay of these international requirements have on U.S. mariners?

Response: The United States, in alignment with the global maritime community, voluntarily agreed to take a pragmatic approach to enforcement of certain provisions in the recent amendments to the International Convention on Standards of Training,

Certification and Watchkeeping for Seafarers (STCW), during the period between January I and July **1,** 2017. The Coast Guard participated in a global effort at the International Maritime Organization to develop guidance tor member states to recognize that many countries were facing the challenge of processing a large number of applications from mariners who waited until the final weeks before the January 1 implementation date to submit their applications. The relatively late submission of these applications caused backlogs in the credentialing evaluation processes of a number of countries, including the U.S. Factoring this information into enforcement action helped to reduce the chances of detention of U.S. vessels trading internationally and foreign flagged vessels visiting U.S. ports. The economic repercussions of wholesale detentions for STCW-related issues would have been significant to the maritime industry as well as U.S. consumers.

In advance of the implementation deadline, the Coast Guard published several documents, in which we encouraged mariners to continue their pursuit of compliance with the STCW requirements.

We believe there were no detrimental impacts to U.S. mariners, vessels, or safe commerce brought on by this pragmatic approach to enforcement.

Question#:	5
Topic:	WAK COTP Available Equipment
Hearing:	Authorization of Coast Guard and Maritime Transportation Programs
Primary:	The Honorable Don Young
Committee:	TRANSPORTATION (HOUSE)

Question: Does the current level of equipment available in the WAK COTP zone meet the needs of the Coast Guard?

Response: The standards tor oil spill response capability and capacity are established in the National Planning Criteria. In cases where the National Planning Criteria cannot be met due to current resource and infrastructure challenges (as are found in Western Alaska), the Coast Guard may accept Alternate Planning Criteria as a risk mitigation measure until the National Planning Criteria can be met. The current level of response equipment in Western Alaska does not yet meet the National Planning Criteria.

Question: If not, how will the Coast Guard achieve the level of equipment it believes is needed in the WAK COTP zone? Docs the Coast Guard expect a steady increase of equipment and response capability?

Response: Yes, the Coast Guard expects industry to steadily increase response capability over time. This was, and remains, the clear intent of the VRP regulations. The Coast Guard will continue to facilitate and foster continued growth toward the national planning criteria.

Question#:	6
Topic:	Equipment Build-out
Hearing:	Authorization of Coast Guard and Maritime Transportation Programs
Primary:	The Honorable Don Young
Committee:	TRANSPORTATION (HOUSE)

Question: Do you believe that equipment build-out in the WAK COTP zone is trending in the right direction?

Response: Yes, the growth of response capacity within the Western AK Captain of the Port Zone over the last 18+ months is positive. The additional resources have not only added to oil spill response readiness, but have also been vital to other missions including search and rescue, and salvage responses.

Question#:	7
Topic:	Equipment Measurement
Hearing:	Authorization of Coast Guard and Maritime Transportation Programs
Primary:	The Honorable Don Young
Committee:	TRANSPORTATION (HOUSE)

Question: How does the Coast Guard measure "equipment"? What are the current levels of equipment in the WAK COTP Zone?

Response: The Coast Guard measures equipment and the planned ability to respond in accordance with the regulations, which include specific levels of response equipment and oversight based on vessel service, the type of oil carried, and the worst case discharge. This applies in Western Alaska as it does in other parts of the country.

Question: How is ability to deploy equipment taken into account when measuring the level of equipment available?

Response: The Coast Guard measures equipment and the planned ability to respond in accordance with the regulations, which include specific levels of response equipment and oversight based on vessel service, the type of oil carried, and the worst case discharge. This applies in Western Alaska as it does in other parts of the country.

For oil spill response equipment, the Coast Guard reviews contracts, funding agreements, and certification statements between plan-holders and their contracted Oil Spill Removal Organization(s) (OSRO) stating their capability to respond to a worst case discharge in the COTP zones they operate. In addition, the Coast Guard measures the oil spill response equipment through an OSRO classification system managed by the National Strike Force Coordination Center. Traditionally, OSROs have voluntarily participated in the OSRO Classification system, which includes on-site verification of equipment by Coast Guard inspectors.

Question#:	8
Topic:	Ships Transiting the WAK COTP Zone
Hearing:	Authorization of Coast Guard and Maritime Transportation Programs
Primary:	The Honorable Don Young
Committee:	TRANSPORTATION (HOUSE)

Question: Are the number of ships transiting the WAK COTP Zone increasing? How many of those ships are considered tank vessels as compared to nontank vessels?

Response: The number of transits across the great circle route grew from 2015 to 2016; from 7735 to 8680 transits. Of these transits, approximately 7% were made by tank ships. These numbers do not include fishing vessels or intra-Alaska tug/barge traffic. Of the above, approximately 15% of transits are ships on innocent passage. Ships on innocent passage are not required to maintain USCG-approved Vessel response Plans.

Question#:	9
Topic:	Revenue Available
Hearing:	Authorization of Coast Guard and Maritime Transportation Programs
Primary:	The Honorable Don Young
Committee:	TRANSPORTATION (HOUSE)

Question: Is the Coast Guard tracking the overall level of revenue available in the WAK COTP zone for compliance with APC rules? If the revenue generated decreases, will the expected amount of equipment and level of response decrease?

Response: No, the Coast Guard does not track data related to level of available revenue. The Alternative Planning Criteria policy has resulted in additional resources in Western Alaska. These resources have not only added to the response readiness of Western Alaska for oil spill response, but have also been vital to other missions including search and rescue responses, and salvage responses.

Question#:	10
Topic:	Marine Safety Information Bulletin
Hearing:	Authorization of Coast Guard and Maritime Transportation Programs
Primary:	The Honorable Don Young
Committee:	TRANSPORTATION (HOUSE)

Question: Are all APCs in the WAK COTP Zone required to comply with the current Marine Safety Information Bulletin and other regulatory requirements issued by the Coast Guard?

Response: Yes, all APCs in WAK COTP Zone are evaluated against the guidance provided in the current MSIB and other regulatory requirements, including Policy Letter 09-02. Once signed, the updated APC national policy guidance will replace Policy Letter 09-02.

Question#:	11
Topic:	Coverage Requirements
Hearing:	Authorization of Coast Guard and Maritime Transportation Programs
Primary:	The Honorable Don Young
Committee:	TRANSPORTATION (HOUSE)

Question: Do the current APCs cover all of the WAK COTP Zone? If not, what is the requirement for coverage'? Docs it vary from APC to APC and who ultimately decides the extent of coverage?

Response: No, the current APCs do not cover all of WAK COTP Zone. Per the regulations, the requirement for coverage is determined by the vessel owner/operator as an area within which the National Planning Criteria are inappropriate and the vessel intends to operate. The Coast Guard evaluates each submission to ensure that the national planning criteria are, in fact, inappropriate for the areas that the vessel intends to operate.

Question#:	12
Topic:	Permanently Positioned
Hearing:	Authorization of Coast Guard and Maritime Transportation Programs
Primary:	The Honorable Don Young
Committee:	TRANSPORTATION (HOUSE)

Question: Does the Coast Guard require that all equipment provided by approved APC providers be permanently positioned in the WAK COTP? If not, and if current APC providers reference equipment not located in the WAK COTP, can you tell me where that equipment is currently located and how it will be made available in timely manner to respond to a spill?

Response: The Coast Guard evaluates each APC on its own merits, including its risk reduction and proposed gap mitigation strategies. In the case of the Alaskan APCs, those strategies all include a significant amount of response equipment located in the covered areas as well as a cascade proposal from elsewhere in Alaska and the lower 48.

Similarly, oil spill response equipment in the lower 48 is based, to a certain extent, on the ability to cascade equipment from other locations.

STATEMENT OF ACTING CHAIRMAN MICHAEL A. KHOURI FEDERAL MARITIME COMMISSION, BEFORE THE COMMITTEE ON TRANSPORTATION AND INFRASTRUCTURE, SUBCOMMITTEE ON COAST GUARD AND MARITIME TRANSPORTATION, UNITED STATES HOUSE OF REPRESENTATIVES, APRIL4, 2017

Chairman Hunter, Ranking Member Garamendi, and Members of the Subcommittee, thank you for providing me with this opportunity to appear before you today and discuss the mission of the Federal Maritime Commission.

The Federal Maritime Commission is working to be a more efficient organization by making a concerted effort to reduce regulatory burdens on our constituents. The Commission is aggressively looking for ways to make compliance with Commission requirements easier and more cost effective for shippers, carriers, and ocean transportation intermediaries. On March 6th, the Commission approved key changes to regulatory requirements for ocean carrier service contract filings and non-vessel-operating common carrier (NVOCC) service arrangement filings that will make it easier and more efficient for shippers and carriers to do business.

I want to assure you, and those with business before the Federal Maritime Commission, that I am committed to continuing to identify rules that arc outdated or impede the efficient operation of business, and eliminating them whenever possible. I believe there are additional reforms that can be made to existing regulations administered by the Federal Maritime Commission, and relief from tariff publication requirements immediately comes to mind as another obligation ripe for Commission consideration.

Currently, ocean transportation intermediaries and vessel-operating common carriers are required to publish tariffs, even though the overwhelming majority of cargo moving across oceans docs so under the terms of service contracts. In other words, current law and Commission regulations require companies to publish rates that have nothing to do with the actual market prices being charged to shippers. Continuing to mandate thousands of tariffs be published that do not ret1ect real conditions in the market, and have minimal, if any, use by industry participants when negotiating service contracts, is a requirement and expense that regulated entities could be relieved of under the exemption authority provided to the Commission by Congress.

Building on Executive Orders of both the prior and current administrations, the Commission is in the process of identifying requirements that should be changed or eliminated. It takes far too long to travel the road to regulatory relief. When we see a regulatory burden, we should be able to successfully address it as rapidly as possible, and by that, I have in mind a timeframe of months as opposed to years. Unshackling the market from artificial constraints on efficiencies should be something we designate as a priority and work tirelessly to achieve.

Toward that end, after consulting with my fellow Commissioners, on March 13th I designated the agency's Managing Director, Karen V. Gregory, as the Regulatory Reform Officer. Ms. Gregory will stand up a Regulatory Ref01m Task Force as contemplated by the January 301 “Presidential Executive Order on Reducing Regulations and Controlling Regulatory Costs.” The task force will identify those regulations that arc the most ineffective, would be the easiest to repeal, and then establish a definitive timeline within the agency to move those items to a vote before the Commission. Even if not technically required as an independent agency to take this step, I believe that it is the right action to take and is consistent with the broader deregulatory history and scope of the Shipping Act.

This commitment to deregulation is essential. It is a critical factor in expanding America's economic competitiveness in the ocean container supply chain. If I can achieve only one thing as the Acting Chairman of the Federal Maritime Commission, I would like our numerous stakeholders to view the agency's regulatory regime as fair and to view the FMC as a global leader in terms of promoting a competitive and efficient ocean transportation system, with primary reliance on a free and open market place, and a minimum of government intervention and regulatory costs.

While my enthusiasm for achieving meaningful deregulation is genuine, no one should mistake that priority as signaling decreased vigilance in meeting the Commission's core mission-fostering a fair, efficient, and reliable international ocean transportation system while protecting the public from unfair and deceptive practices. We have been doing this work, as one entity or another, for one hundred years and we are at a particular juncture in the history of the container shipping industry where our work is all the more important.

Since the Commission last appeared before you, there have been tremendous changes to the ocean transportation services marketplace. The recent past has been marked by considerable merger and acquisition activity among shipping lines that was topped off by the bankruptcy of a "top ten" carrier late last summer. As a result of these events, the number of major shipping lines operating in the international trades has dropped from 20 in 2015 to what will be 13 by next year when the three Japanese-based carriers create a new, consolidated container line. All of this activity has consequentially caused a reordering of the carrier alliance system and the creation of two new organizations (“THE Alliance” and “The OCEAN Alliance”) that will join the already existing “2M Alliance.”

Carrier and marine terminal alliances can be very beneficial for U.S. exporters, importers, and consumers. Such alliances allow participants to obtain efficiencies and cost-savings that can be passed on to domestic consumers especially when healthy competition exists among vessel operators. Last, the benefits of alliances and other forms of joint commercial arrangements arc recognized and addressed in the Shipping Act of 1984, as amended, and the contemporaneous Congressional record.

Given the significant amount of change that has taken place over the past year, shippers are viewing the new commercial environment with many questions and perhaps even some trepidation. Put succinctly, shippers are concerned that fewer carriers, operating in fewer, yet larger, alliances will not only equate to less service choices, but to a commercial environment where shipping lines enjoy an advantage of leverage in contract negotiations. These apprehensions are not limited exclusively to cargo owners. We have heard similar sentiments voiced by tern1inal operators, equipment lessors, as well as various service providers. The industry is entering a new era and it is not surprising that there may be questions about whether carriers will be in a position to dictate rates. The current circumstances perfectly illustrate why the Federal Maritime Commission was created, what its job is, and how the agency provides a benefit to American shippers and to our economy more broadly.

The mission of the Federal Maritime Commission is to facilitate an open and free market for ocean shipping services by bringing transparency to market forces and protecting against anticompetitive behaviors. While the Commission is always diligent in its duties to assure the marketplace remains competitive, we are aware and alert that with the dawn of a new era, the oversight work of our professional transportation economists, attorneys, and analysts is all the more critical. It is vital to the interests of all parties involved in the movement of ocean commerce that thc Commission provide confidence that America has a competitive, fair, and efficient marketplace. As the new alliances begin to operate commercially next month, we are positioned to monitor the marketplace more carefully than ever with a particularly keen eye focused on any behavior that might appear to produce an unreasonable increase in transportation cost or an unreasonable reduction in transportation services as a result of collective carrier behavior under the respective alliance agreements. Toward that goal, thc Cmm11ission has strengthened the economic review process of new alliance agreements filed with us. More specifically, we have required tighter limits on the scope of each agreement's authority, as well as expanded quarterly reporting requirements to be filed with the Commission. All of this is done to ensure the Commission's ability to detect and respond to any signs of anticompetitive behavior or abnormal pricing trends.

Our *Annual Report* will provide you with a comprehensive summary of the Commission's activities and industry developments in Fiscal Yem·2016. While future predictions are difficult, 1 will briefly address what we foresee as potential developments and trends in the coming months.

Carriers are operating in an environment where trade volumes are growing more sluggishly than has been typical of the sector. Concurrently, new vessels of the largest capacities continue to emerge from shipyards and be deployed into the world's trade lanes. All things being equal, and with no other shocks to the business such as a spike in fuel costs or another shipping line bankruptcy, this is a formula for continued low rates, which ultimately benefits shippers, at least in the shorter term.

As noted earlier, although there has been a contraction in the number of lines operating in the international ocean trades, competition between companies remains vibrant and shippers continue to benefit from low rates. Overall market share of even the largest oceangoing carriers remain diffused. In the U.S. export and import trades combined, CMA CGM holds a 12.42 percent market share followed closely by Mediterranean Shipping Company at 12.39 percent and Maersk in third position with 10.62 percent. These are far from dominant market positions as recognized by established economic standards. We will, of course, monitor what impact the carriers operating in the new alliances have on market dynamics, rates, and services. A preliminary analysis conducted by FMC staff, and hopefully available for publication later this year, finds that shippers may enjoy certain benefits from larger alliances, particularly in tcm1s of what they pay for service.

One factor to note is that some carriers receive government support, either directly or indirectly. The invisible hand is not the only force that guides the global shipping industry, and nations throughout the world go to great lengths to support national companies, including saving them from bankruptcy. At the moment such close links between a govcrru1 1ent and its national carrier can also benefit American shippers and consumers by virtue of lower freight costs and greater service choices.

The ocean liner industry has been in a state of vessel oversupply for several years. The low freight rate structure in U.S. trade lanes is a direct reflection of that capacity supply/demand imbalance and American exporters and importers have been the beneficiary of those low freight rates. Such supply imbalances will not last forever. The Federal Maritime Commission does not favor one competitor, sector, or industry stakeholder over another. We will continue to be vigilant for indications of rate increases that are products of market distorting, collusive carrier business practices; however, it is important to remember that rate increases in and of themselves are not proof of an uncompetitive marketplace. At some point in the future, higher freight rates will be a normal result of a more equalized and healthy supply/demand marketplace.

Finally, I should note that the Commission is beginning to see marine terminal operators and port authorities show increased interest in how to use agreements filed with and reviewed by the Commission to their benefit. There is a realization among these parties that seeking an alternate antitrust enforcement regime available to them through an agreement filed at the FMC can lead to increased efficiencies and lowered costs. We welcome the application of any parties from the port and terminal sector who want to use agreements to achieve goals that ultimately benefit the American shipper and consumer.

Turning from the shipping industry to the Commission itself, I would like to note an FMC initiative that is making a meaningful contribution toward understanding, addressing, and mitigating port congestion- as well as enhancing supply chain efficiency for America's exporters and importers. I am referring to our Supply Chain Innovation

Team Initiative that is being led by my colleague and friend, Commissioner Rebecca Dye.

This undertaking was launched in May of 2016 and focuses on challenges faced by America's international maritime supply chains. Commissioner Dye, with her volunteer teams of industry leaders composed of shippers, marine terminal operators, trucking companies, ocean carriers, port officials, labor representatives, logistics companies, and other stakeholders, have been working to develop actionable commercial solutions including in particular-· the key content for a national seaport information pm1al that could provide the necessary critical inf01111ation sought by all parties involved in moving containers to/from vessels, through seaports, and onward to a final destination.

The team's consensus in Phase One was that if such a portal could be funded and implemented, it would make a meaningful contribution to both improved supply chain efficiency and to America's national economic competitiveness.

Phase One of the initiative addressed our country's import supply chains. This year, Commissioner Dye will launch a second phase this time focusing on America's export supply chains. We are very much looking forward to the results of the work those teams will do.

Commissioner Dye's work illustrates what can be accomplished with the right people working toward a common goal under effective leadership. A proven method of achieving strong performance at an organizational level is through focused and meaningful strategic planning. From my years in the private maritime sector, I experienced first-hand how strategic planning can be not just a tool, but a driving force in an organization's success. Government agencies can also benefit from strategic planning that is focused, and designed to unite all agency team members to find ways to achieve our mission more effectively while delivering value to the taxpayer. Creating a strategic plan that is substantive and seeks to make the Commission an even better run agency is a task I welcome taking on as the Acting Chairman. To this point we are now beginning work on a new five year strategic plan that will guide the Commission through Fiscal Y car 2022.

Finding ways to conduct the Commission's business more efficiently is an important goal we share, Mr. Chai1man. I am very pleased to say we have made some definitive steps toward trying to find ways to make every dollar appropriated to us go as far as it can. The FMC is a small agency with a very technical mission and a need for a very specialized workforce. Our requested level of funding for Fiscal Year 2017 is $27,490,000. The part of the agency involved in the heart of the mission, you will find a high percentage of transportation economists and attorneys – career fields that tend to command more compensation in order to successfully recruit and retain qualified candidates. Overall, the great bulk of the Commission's budget, approximately 88 percent, is consumed by rent and salaries. "Overhead" costs such as security fees, utilities, and government services account for most of the remaining budget dollars. The

Commission has very little, if any, control over many of these costs. Year in and year out, the rent we are charged rises, the supplies and resources we purchase to support our economist and attorney's competitive analysis and legal research cost more, and information technology costs – including IT security and telecommunications bills – rise. We constantly work to find a balance between our resources and our workload; however, if there is a surge of agreement filings, if a "class" of plaintiffs choose to seek relief at the FMC, or if our building security requirements increase, then we work to prioritize our mission-critical activity and reallocate resources to the extent possible.

Another example of the challenge of controlling our costs is illustrated by the ongoing discussions we are having with the General Services Administration (GSA) regarding security requirements and related costs for our headquarters building in Washington D.C. The Commission's projected security costs for FY2017 is $758,017. Of this total, only $71,983 can be categorized as being basic, necessary, and pertinent for the agency's headquarters building and its six regional Area Representative otlices. The additional $686,034 we estimate we will pay in security fees are incremental costs imposed upon us because of the building where are our headquarters is located. The $686,034 figure represents approximately 2.67 percent of our total FY 2017 budget. For an agency with a low security requirement and a desire to be accessible to the public, that money could be better used to hire the additional economists and analysts we need to help monitor the international ocean shipping marketplace.

Our building is a mix of government and non-government tenants. Two of our fellow government tenants are law enforcement agencies with higher physical security requirements than the Federal Maritime Commission. We are concerned that we will be required to shoulder anticipated cost increases associated with maintaining security levels at the building that are dictated by the needs of other federal tenants co-located by GSA into our building subsequent to our original occupancy in 1991. Additionally, we are concerned that we will be directed to contribute toward the overall cost of new security upgrades to meet the desires or professed needs of other federal tenants in the building. These ever-increasing security costs are problematic in light of current budget realities.

As I mentioned earlier in this testimony, the Federal Maritime Commission continues to faithfully pursue the purposes and mission of the Shipping Act that has been in place for I 00 years. In its *First Annual Report,* published on December 1, 1917, a total of 83 employees were listed on the payroll of our predecessor agency, the United States Shipping Board. Today, a century later, with a vastly larger domestic economy and global trade, the Commission has only 127 employees working to maintain a competitive ocean transportation services marketplace that moves billions of dollars in commerce representing about one-third of America's Gross Domestic Product. I am proud of the work the Commission's staff does each day to ensure a fair, efficient, and reliable international ocean transportation system, and to protect our U.S. exporters, importers, and ultimately the American consumer from unfair and deceptive practices.

Thank you for this opportunity to discuss the mission of the Federal Maritime Commission, current state and future challenges of the ocean shipping industry, as well as highlight some of the Commission's recent achievements and future priorities. I am always ready to be of any assistance to the Subcommittee and its Members. I would be pleased to answer any questions you may have.

RESPONSE OF ACTING CHAIRMAN MICHAEL A. KHOURI TO QUESTIONS FOR THE RECORD (QFR) SUBMITTED ON BEHALF OF CONGRESSMAN DUNCAN HUNTER

Why did the FMC agree to allow a group of international car carriers to collectively negotiate wit/t U.S. tugboat service providers?

Because attempting to stop the agreement in Federal District Court as a violation of the Shipping Act of 1984 would have been frivolous, and would have likely exposed the Federal Maritime Commission ("FMC" or "Commission") and its attorneys to sanctions under Rule II of the Federal Rules of Civil Procedure.

When the Commission receives an agreement, several things happen. First, staff conducts a preliminary review to determine that the parties to the agreement are vessel operators or marine terminal operators, and that there are no overt prohibited acts on the face of the agreement. Next, after being satisfied that there are no facial defects, staff then conducts a competitive analysis using well-worn antitrust standards. The analysis defines the market-both product and geographic. For agreements with joint purchasing authority, staff examines whether the portion of goods or services that would be jointly purchased is so large, relative to total purchases of the good or service, that the parties could effectively exercise market power in the upstream market.

In the case of the car carrier agreement amendment, the market being evaluated was the market for tug services for ocean going vessels that require tug assist in port docking. The geographic market was each relevant major port where the car carriers called. The FMC staff analyzed 12 U.S. ports on the west, east and gulf coasts. The basic assessment was that the car carriers did not have significant market purchasing power. Just as DOJ and FTC assess purchasing joint ventures under a rule of reason standard (testing whether the cooperative activities may produce efficiencies), the Commission assessed whether there would be efficiencies and cost savings that might flow to consumers. Unlike arrangements between competing sellers or between the competing service providers, which are judged skeptically using a *per se* test for illegality, joint ventures for purchasing are permitted - and even encouraged.

Here, car carriers make up a comparatively small part of the entities that purchase tug assist services. It thus would have been beyond difficult to make the legal case, based on the present facts that the agreement would have caused an unreasonable reduction in transportation service, or an unreasonable increase in transportation costs (the standard that the Commission would have been required to meet in order to convince a judge to grant an injunction).

When it determined that the legal case would have been fruitless, the Commission nevertheless implemented safeguards to ensure that the car carriers' agreement would not result in unexpected consequences. Specifically, the Commission notified the car carriers that before implementing any actual joint purchase, the car carriers would be required to bring *that actual proposed tug assist agreement* back to the Commission for review in order to ensure that the car carriers did not exercise any undue power in a particular port. Failure to do so would result in an investigation and potential enforcement proceeding. Further, the car carriers *do not receive antitrust immunity* for agreements with tug providers. *See* 46 U.S.C. § 40307(b)(l). Additionally, the tug providers can file a complaint for adjudication at the Commission based on prohibited acts. *See* 46 U.S.C. § 41105(4).

In sum, the Commission did not seek to enjoin the agreement because the parties are unlikely to be able to exercise market power in the tug services market, the Commission would have lost in court with further risk of court-imposed sanctions on the FMC, and the tug providers still have several adequate, currently existing remedies if they do not concur with the decision of the Commission.

Given the increased consolidation we /lave seen in tile foreign ocean carrier industry over tile past decade, do you think the Shipping Act adequately protects the U.S. maritime industry and tile public from the concentrated power of foreign shipping alliances?

As I noted in my written testimony submitted for the May 3 hearing before the Subcommittee on Coast Guard and Marine Transportation, the Commission has ample authority under the Shipping Act of 1984 (Shipping Act) to address the competitive issues facing the international liner industry. As an independent agency with experience and specialized expertise in this legal area and particular familiarity with maritime matters, the FMC is well positioned to be sensitive to the concerns of U.S. participants in the international ocean transportation supply chain and the U.S. shippers who rely on a fair, efficient, and competitive, ocean transportation system. The FMC has been forward leaning in our use of our authority as we have reviewed new carrier agreements as they formed the new generation of alliances. We have also been careful to consider the concerns of parties affected by these agreements, including the U.S. maritime industry and shipping public, as well as the views of our sister competition agencies.

Under the Shipping Act, cooperative or collaborative agreements between or among competitor international ocean liner carriers are filed with the Commission and reviewed under the Shipping Act's competition standard to prevent anticompetitive behavior in these agreements. This standard the Commission uses to review carrier agreements, 46

U.S.C. §41307(B)(l) - "Anticompetitive Agreements," commonly referred to as 6(g), is analogous to the standard employed by the Department of Justice (DOJ) and the Federal Trade Commission (FTC) to review mergers, acquisitions, and competitor collaborations. Under 6(g), an agreement filed with the Commission goes into effect *unless* the Commission determines (and convinces a judge to agree) that the agreement *is likely, by a reduction in competition, to produce an unreasonable reduction in transportation service or an unreasonable increase in transportation cost.* In the event of such determination, the Commission then must go to a Federal District Judge as discussed below. Drafted by the House Judiciary Committee as a part of the deliberations leading to enactment of the Shipping Act, the process for agreement review under 6(g) is modeled on the Hart-Scott-Rodino Act of 1976 governing premerger clearance of proposed acquisitions and mergers. Congress adapted this process for the Commission as part of the Shipping Act to ensure that that potential efficiencies and cost-savings would not be lost by consumers because of delay in agreement effective dates. In order to prevent the agreement from going into effect, the Commission must bring a civil action in the United States District Court for the District of Columbia and successfully obtain an injunction to halt the operation of the agreement. The burden of proof is on the Commission.

If parties agree to undertake activities that are governed by the Shipping Act, but do not comply with the Commission's process of review, they risk not only Shipping Act sanctions, but also federal criminal sanctions prosecuted by DOJ under the Sherman Act.

The success of section 6(g) in allowing the Commission to address concerns about potentially anticompetitive effects of filed agreements is evident in the Commission's successful efforts to mitigate or eliminate potentially anticompetitive provisions in pending agreements through detailed discussions with filing parties during the review process. By its terms, the Shipping Act provides an opportunity for the public, including U.S. maritime interests, to express concerns about filed agreements. The Commission takes these comments seriously, and uses them, together with its own economic analysis under 6(g), during the review process and in its ongoing review of trade conditions and carrier behavior under filed agreements to consider and address anticompetitive concerns.

During the period of time following the filing of the 2M, OCEAN, and THE alliance agreements and the time they went into effect, Commission staff engaged in detailed discussions with the filing parties to understand the scope and ramifications of proposed agreement authorities. Of particular concern were the parties' ability to jointly contract with vendors and suppliers, the possibility of exchanging commercially sensitive information and any potential limitation on the ability of individual alliance carriers to make independent decisions regarding their vessel capacity. As a result of these discussions, the alliance parties for the 2M, OCEAN, and THE alliances submitted substitute language that either eliminated or mitigated these concerns. For example, 2M does not allow for joint contracting. Under THE Alliance and the OCEAN Alliance joint contracting for tug services would not take place in the United States; joint contracting with marine terminal operators will take place only if the marine terminal operator wishes to engage in joint negotiations, and the type of commercial information that can be exchanged between the Parties was clarified and narrowed.

In addition to the review of carrier agreements for potentially anticompetitive effects under 6(g), the Commission may use section 10, the "Prohibited Acts" provisions in the Shipping Act, to preserve competition. This section of the Shipping Act includes prohibitions on a number of business practices on concerted carrier conduct acting outside of approved authority (such as price fixing or market allocation), unreasonable practices, discrimination in price or accommodations, refusal to deal, retaliation, boycotts, predatory practices, and discrimination based on shipper affiliation. 46 U.S. Code § 41105(4), prohibits carriers from jointly negotiating with non-ocean carriers *if doing so would violate antitrust laws* (emphasis added). These prohibited practices mirror remedies found in other competition statutes, such as the Robinson-Patman Act of 1936.

The Commission, of course, may enforce section I 0; but private litigants may also bring actions under these Shipping Act provisions to protect their interests.

The international ocean liner industry has recently experienced some consolidation because of long-term structural issues in the industry and poor financial returns and the number of major carrier alliances serving the U.S. trades decreased from four to three. The reduced number and increased size of these major alliances have raised new issues and concerns for the FMC and changed the way in which the Commission approaches these joint ventures. Broader authorities and language for small or limited slot sharing agreements or in a world with seven or eight alliances with much smaller market shares presented fewer and less complex competitive issues. Provisions that might have been acceptable in earlier agreements for smaller and more limited joint ventures have become increasingly problematic as the number of alliances serving the U.S. trades has shrunk to four, and now three. The FMC has increasingly insisted on narrower authorities, more specific language, and enhanced monitoring requirements. For some time now with respect to these larger alliances, the FMC has required more "clear and definite" authority language for alliances. Monitoring for these large alliances, requiring more details and more frequent filing of monitor reports has increased.

Each of the relevant geographic and product markets in which an agreement or alliance would operate is reviewed and evaluated by the Commission using standard antitrust economic principles. This includes assessing whether the agreement or alliance parties would have the potential and the incentive to use unilateral or coordinated actions that may restrain competition in any of these markets. A large, high-volume, high-growth, market with many participants and numerous potential entrants ("contestability") may not be problematic, for example, but a low- volume, low-growth, distant market with few current participants may raise competition concerns. When necessary, the Commission also evaluates whether the liner trade between any given foreign country and a specific U.S. coast (e.g., between Japan and the United States west coast) is its own separate market or is part of a broader geographic market.

Because alliances are ongoing cooperative agreements rather than mergers, the Commission is charged by Congress with ongoing and continuous monitoring after the initial review and following the effective date of the agreements. The Commission checks for anticompetitive behavior that would violate the Shipping Act. The Commission may challenge an agreement at any time after the effective date. Our expert analysts, economists, and attorneys maintain a careful watch on industry trends, being vigilant for any indications of anticompetitive behavior by the participants operating within the filed agreements. As a general rule, subject to the competition standard of section 6(g) and the prohibited acts in section I 0, the Shipping Act permits joint ventures among ocean carriers, these vessel sharing agreements can create efficiencies and cost savings that are passed on to exporters, importers, and ultimately benefit the U.S. consumer. Historical data points indicate that cost saving initiatives are being passed over to shippers. For example, between 1998 and 2013, average container freight rates, as measured by the China Container Freight Index increased by three per cent. Yet, over the same fifteen year period, fuel price has increased by 790 per cent. According to Alphaliner, in 1998, fuel was eight per cent of total container freight costs. In 2013, it accounted for over twenty-five per cent of total freight cost. It is widely reported in the trade press that carriers are in a constant battle to find cost savings. This figure would be considerably higher but for the practice of slow-steaming, which is one tactic widely used by carriers in their constant battle to find operational cost savings.

Under the 2M, OCEAN, and THE alliance agreements, the parties are permitted to jointly contribute and share vessels, charter or swap space on each other's vessels, and engage in a variety of cooperative working arrangements pertaining to operating services and related equipment. The latter would include, for example, deciding the services to be deployed and their associated schedules, coordinating sailings, establishing service performance criteria, agreeing on vessel maintenance and repair schedules, and the substitution of vessels during dry-docking. The alliance parties may exchange information pertaining to vessel operations, market conditions, and third- party costs. They also may establish an operations coordination center to manage and monitor the service network to ensure schedule integrity and service quality. Services within the network are reviewed at least annually, and adjusted where necessary, to reflect each member's respective requirements for the movement of cargo. Moreover, each alliance permits its members to charter space from (or sell space to) non-alliance parties if needed. Parties to the OCEAN and THE alliances are even allowed to deploy additional services in alliance trade lanes (but not under the alliance agreement) provided they give their partners first right of refusal of any new capacity offered.

Under the agreements as they went into effect, participants in these three alliances may not jointly contract with tug operators or other vendors of vessel-related goods and services in the United States. The parties will negotiate independently with marine terminal operators, except where the marine terminal operator is agreeable to a joint contract with the parties, in which case a joint contract with the marine terminal operator would be authorized, and subsequently filed with the Commission.

These alliance parties do not have authority to discuss and agree on rates or to exchange information on freight rates, tariff items, service contract terms or bids, customer lists, individual marketing plans, or similar commercial matters. Therefore, within each alliance and among the three alliances, there continues to be intense rivalry to obtain cargo through vigorous price competition and quality service delivery.

The Commission applies the Shipping Act's 6(g) competition standard to requests for joint purchasing authority just as it does for agreements seeking joint operational authority. The larger the market share of the participants to an agreement, the more concern the Commission will have about potential "market power" and potentially anticompetitive effects. Whether the Commission or any other antitrust enforcement agency will allow this joint purchasing authority to remain in an agreement depends on the specific facts of the agreement and an analysis of the potentially anticompetitive effects in each relevant geographic and product market The Commission's determination of the competitive impact of an agreement is based on the specific facts of the requested authority. Such review would be very similar, if not identical to the review by the Department of Justice for joint purchasing arrangements. If anything, the Commission's review of joint purchasing arrangements is more focused than its antitrust sister agencies. While DOJ might permit joint purchasing after a survey of a marketplace, the FMC not only surveys the general marketplace, but also looks at each individual agreement to assess the relevant product and geographic market at the time of the proposed procurement. For joint purchasing collaborations by competitors, the Commission does not use a "safe harbor" formula employed by DOJ and FTC. Employing its industry expertise, the Commission analyzes each agreement on a case-by-case basis, requiring that contracts entered into pursuant to any agreement negotiating authority be brought back to the Commission for further review based on the specific facts.

It is important to note that ocean carrier agreements filed with the Commission do not exempt them from application of the general antitrust laws (Sherman Act and Clayton Act) whenever the joint carrier group is dealing with certain domestic businesses, including tug-assist operators. Congress expressly considered these types of agreements and explicitly excluded these arrangements from the types of joint ocean carrier agreement that are not subject to general antitrust laws. Because they are cooperative working agreements with two or more ocean common carriers, these carrier joint purchasing agreements must be filed with the FMC for review, however, the agreements nevertheless do not receive immunity from the antitrust laws.[6]

---

[6] This conclusion is reinforced by two statutory provisions:
o 46 U.S. Code§ 40307- Exemption from antitrust laws

In conclusion, under the Shipping Act, the Commission applies accepted antitrust standards and tests to evaluate the potential anticompetitive effects of ocean carrier agreements. The Commission is careful to consider the views and analyze the effects of concerted ocean carrier collaborations on all affected stakeholders. If unable to reach agreement with the filing carriers to remove potentially anticompetitive authorities from an agreement, the Commission will seek an injunction in U.S. District Court to prevent an agreement that allows the exercise of "market power" to the detriment of U.S. participants in the ocean transportation supply chain from going into effect.

Additional statutory authorities added by Congress to the Shipping Act could assist the Commission in gathering necessary and relevant information to evaluate agreements, and could provide additional avenues and remedies, allowing the Commission to better ensure a competitive market-place, protecting the international ocean transportation supply chain, and providing affected private parties with additional avenues to protect their interests. Should Congress determine to amend the Shipping Act, I will commit to implementing and enforcing any change consistent with Congressional intent, just as the Commission had done with the 1984 Act and 1998 amendments.

STATEMENT OF JOEL SZABAT, EXECUTIVE DIRECTOR, MARITIME ADMINISTRATION, U.S. DEPARTMENT OF TRANSPORTATION, BEFORE THE COMMITTEE ON TRANSPORTATION AND INFRASTRUCTURE, SUBCOMMITTEE ON COAST GUARD AND MARITIME TRANSPORTATION, U.S. HOUSE OF REPRESENTATIVES, AUTHORIZATION OF COAST GUARD AND MARITIME TRANSPORTATION PROGRAMS, APRIL 4, 2017

Good afternoon, Chairman Hunter, Ranking Member Garamendi and members of the Subcommittee. I appreciate the opportunity to discuss the Maritime Administration's (MARAD) programs and how they support the agency's statutory mission to foster, promote, and develop the merchant maritime industry of the United States.

{b) Exceptions.-This part *does not extend antitrust immunity* to-

(I) an agreement with or among air carriers, rail carriers, motor carriers, or common carriers by water not subject to this part relating to transportation within the United States;

o 46 U.S. Code § 411 05 - Concerted action

A conference or group of two or more common carriers may not-

(4) negotiate with a non-ocean carrier or group of non-ocean carriers (such as truck, rail, or air operators)

on any matter relating to rates or services provided to ocean common carriers within the United States by those non-ocean carriers, unless the negotiations and any resulting agreements *are not in violation of the antitrust laws* and are consistent with the purposes of this part, except that this paragraph does not prohibit the setting and publishing of a joint through rate by a conference, joint venture, or association of ocean common carriers.

The Maritime Administration (MARAD) implements programs that promote the economic competitiveness, efficiency, and productivity of the U.S. maritime transportation system and ensures sealift capability and capacity is available to support the national and economic security needs of the Nation.

National Security

MARAD traces its origins to the Shipping Act of 1916, which established the U.S. Shipping Board, the first Federal agency tasked with promoting a U.S. Merchant Marine and regulating U.S. commercial shipping. Later, the Reorganization Act of 1950, which codified President Truman's Reorganization Plan No. 21, formally established the MARAD to ensure the maintenance of an adequate Merchant Marine to support national security. By law and (10 U.S.C. Chapter157) and Presidential National Security Directive No. 28 dated October 5, 1989, the Department of Defense (DOD) must rely on U.S.-flag ships crewed by volunteer, civilian American mariners, to provide the sealift to support military deployments and respond to emergencies. MARAD's responsibility is to ensure that U.S.-flag ships and merchant mariners are available to meet DOD requirements. The U.S.-flag fleet of privately owned, commercially operated vessels, along with government-owned vessels, provide a critical public-private sealift surge and sustainment capacity to move equipment and materials for the Armed Forces and Federal agencies when needed, and where needed, during times of conflict, humanitarian crises, and natural disasters.

Three programs ensure that there are enough U.S.-flag vessels available to provide this capacity: the Jones Act, which ensures a role for the U.S.-flag fleet in domestic trade; and Cargo Preference and the Maritime Security Program (MSP) which, together, support a militarily useful, U.S.-flag fleet sailing internationally.

MARAD and DOD also rely on the commercial fleet to employ enough qualified mariners to crew all the commercial ships tasked to support military operations, as well as 1,300 additional mariners to crew the "surge fleet" of Federally-owned cargo ships. As of today, the size and composition of the U.S.-flag commercial fleet is adequate to meet immediate military contingencies. However, due to the decline in size of both the domestic U.S.-flag fleet with unlimited horsepower and unlimited tonnage and the international U.S. flag commercial fleet in recent years, both the U.S. Transportation Command and MARAD are concerned that there are not enough qualified mariners to sustain an activation of the entire sealift fleet, though there has never been a full activation of the entire sealift fleet.

Maritime Security Program

The Maritime Security Act of 1996[7] established the MSP, which ensures access to U.S.-flag ships engaged in ocean-borne foreign commerce and the necessary intermodal logistics capability to move military equipment and supplies during an armed conflict or a national emergency, while currently providing employment for up to 2,400 U.S. merchant mariners each year. Under this program, participating operators are required to commit their ships, crews, and commercial transportation resources upon request by the Secretary of Defense during times of war or national emergency. The MSP provides direct annual stipends for 60 active, commercially viable, militarily useful, privately-owned U.S.-flag vessels and crews operating in international trade. The program is authorized up to $299,997,000 for Fiscal Year (FY) 2017 at an annualized stipend level of approximately $4.9 million per vessel.[8] Of the 81 U.S.-flag vessels that currently trade internationally on a full-time basis, 60 participate in the MSP program.

National Defense Reserve Fleet/Ready Reserve Force[9]

MARAD manages and maintains a fleet of government-owned merchant ships in the National Defense Reserve Fleet (NDRF).[10] This includes 45 Ready Reserve Force (RRF) vessels that are maintained ready for operation within five days for transport of cargo to the area of operation and one RRF off-shore petroleum discharge vessel maintained ready for operation within 10 days to meet critical military requirements. As required by law, our Nation has called upon RRF and NDRF vessels, which include training ships on loan to the six State Maritime Academies (SMAs) and the U.S. Merchant Marine Academy (USMMA), to respond to several disasters, including Hurricane Sandy in 2012 and Hurricane Matthew in 2016. Additionally, RRF and NDRF vessels can be configured to support other emergent situations as was the case in mobilizing the Motor Vessel CAPE RAY for use in the international effort to destroy the Syrian Government's declared chemical weapon stockpile. MARAD completed this mission in August of 2014.

Maritime Training[11]

MARAD provides funding and oversight for mariner training programs to produce highly skilled U.S. Coast Guard (USCG) credentialed officers for the U.S. Merchant Marine. Maintaining an adequate pool of American merchant mariners is vital to both the

[7] Section 2 of the Act created the MSP, but that authorization is now codified and appears at 46 U.S.C. Chapter 531. Authorized funding levels are at 46 U.S.C. § 53111.

[8] The most recent amounts appropriated are found in the Consolidated Appropriations Act of 2016 (P.L. 114-113).

[9] See, 50 U.S.C. §4405 and 46 U.S.C. Chapter 571.

[10] See, 50 U.S.C. § 4405.

[11] The Secretary of Transportation is specifically authorized to provide education and training to U.S. citizens for the safe and efficient operation of the U.S. Merchant Marine in 46 U.S.C. § 51103(a). See also, 46 U.S.C. Subtitle V Part B. See Chapters 511, 513, 515 and 517.

commercial success of the U.S.-flag fleet and to maintaining the capacity needed to project American sea power. The USMMA and SMAs graduate the majority of USCG-credentialed officers, consisting of the highest entry-level merchant marine officers who hold an unlimited tonnage or horsepower endorsement available to crew U.S.-flag ships. These graduates support our Nation as a cadre of well-educated and trained merchant mariners capable of serving in support of military emergency, national emergency, and humanitarian missions.

United States Merchant Marine Academy[12]

The USMMA is an accredited institution of higher education operated by the DOT and managed by MARAD. The USMMA offers a four-year maritime-focused program, centered on rigorous academic and practical technical training that leads to a Bachelor of Science degree, a USCG merchant mariner credential with an unlimited tonnage or horsepower officer endorsement, and, upon application and acceptance, a commission as an officer in the Armed Forces or uniformed services (National Oceanographic and Atmospheric Administration (NOAA) Corps or the Public Health Service (PHS) Corps) of the United States. USMMA graduates incur an obligation to serve five years as a merchant marine officer aboard U.S. documented vessels or on active duty with the U.S. Armed Forces or uniformed services. In addition, they must serve as a commissioned officer in a reserve unit of the U.S. Armed Services for eight years. In 2017, 176 midshipmen are expected to graduate from the Academy.

The USMMA's shipboard training program, or "Sea Year," exposes Midshipmen to life at sea on board commercial and military vessels and provides cost-effective hands-on seamanship and engineering sea time that is required to secure USCG mariner credentials. Midshipmen are required to have 360 days of sea service during their four-year program in order to obtain their USCG merchant mariner credentials. Shipping companies and the U.S. Navy are part of a cooperative effort to ensure that a Midshipman's shore based education is enhanced by the required on-the-job training at sea.

Sea Year is critical to the education and training of Midshipmen at the USMMA, and all training must be conducted in a safe and respectful environment. In the wake of a series of reports that indicated problems with sexual assault and sexual harassment (SASH) and other coercive misconduct, both on campus and at sea, DOT and MARAD leadership suspended commercial Sea Year until we developed a better understanding of the problem and could develop a strategy to ensure the safety of the Midshipmen. To address campus concerns, DOT commissioned a cultural audit of the USMMA and the Superintendent is implementing audit recommendations. In addition to the audit, MARAD brought together a consortium of fourteen leading maritime companies to find

[12] See, 46 U.S.C. Chapter 513.

solutions. Just two weeks after the stand down, the consortium brought forth a proposal to address SASH. MARAD and DOT subsequently created a compliance team that established standards, and collaborated with the consortium and labor to lay out workable criteria for the companies to achieve those standards.

Working collaboratively to resume the program, MARAD, industry, and the unions have adopted comprehensive new policies to ensure that the highest standards for behavior, leadership, and integrity are met. Together we have worked to establish requirements for companies providing Sea Year training opportunities for Midshipmen. To meet these requirements, companies must now show that they have written policies, procedures, and robust training demonstrating zero tolerance for SASH; that they will provide qualified mentors for Midshipmen onboard vessels; and have taken other measures to ensure the safety of Midshipmen. Implementation of these requirements will assist in ensuring both the quality of the anti-SASH programs and each company's accountability when it takes on Midshipmen. At present, four companies have met compliance requirements and resumed hosting Midshipmen on their vessels.

MARAD is committed to doing everything it can to eliminate SASH at the USMMA, improving the reporting rate, taking appropriate action in each reported case, and ensuring Midshipmen subjected to assault or harassment receive the services they need. In addition to the efforts to improve the Sea Year training experience, the USMMA has developed a comprehensive plan to reduce SASH on campus. The USMMA Sexual Assault Prevention and Response (SAPR) Program has significantly improved training across the Academy aimed at the prevention SASH, including online prevention training, case studies, videos, social media, professional speakers and small groups. Actions taken by the USMMA have included installation of new emergency call boxes and security cameras, improvement of the security guard force, implementation of a 24/7 hotline for reporting inappropriate behaviors, and victim assistance in obtaining medical or mental health treatment. Efforts will continue to improve upon the SAPR Program as the USMMA implements recommendations from the cultural audit and responds to feedback from Midshipmen.

State Maritime Academies[13]

In addition to providing oversight of the USMMA, MARAD provides funding assistance to six State Maritime Academies (SMAs), which collectively graduate more than two-thirds of the entry-level Merchant Marine officers annually.[14] Approximately 972 Cadets are expected to graduate from the SMAs in 2017. MARAD provides annual direct payments to provide for operational support to each of the six SMAs as well as

[13] See, 46 U.S.C. Chapter 515.

[14] The six SMAs are: California Maritime Academy in Vallejo, CA; Great Lakes Maritime Academy in Traverse City, MI; Texas A&M Maritime Academy in Galveston, TX; Maine Maritime Academy in Castine, ME; Massachusetts Maritime Academy in Buzzards Bay, MA; and State University of New York (SUNY) Maritime College in the Bronx, NY.

assistance to fund the enrollment of 75 new cadets in the Student Incentive Program (SIP) for a period of four years. The SIP program provides cadets with funds to be used for uniforms, tuition, books, and subsistence. Upon graduation SIP students must maintain an unlimited USCG credential for six years, fulfill a three-year service obligation, and serve in a reserve unit of an Armed Forces or uniformed service for eight years. Assistance provided to the SMAs also includes funding for maintenance and repair costs for training ships on loan from MARAD and funding to offset training vessel fuel costs. Unlike the USMMA Midshipmen, the SMA Cadets receive most of their sea time on these training ships, under the instruction of each school's faculty.

The SMAs use MARAD funds for maintenance and repair projects. This work is particularly important as the training ships age and approach or exceed their designed service life. Two training ships are over 50 years in service, which is twice the standard service life. Accordingly, MARAD is using the funds to address priority maintenance across all the training vessels, with emphasis on the Training Ship EMPIRE STATE, to ensure that they all meet safety and functional requirements and remain in service as long as necessary. As required by the Consolidated Appropriations Act, 2016, P.L. 114-113, MARAD has done a standard training ship design and is conducting an independent requirements and alternatives analysis of Cadet training ship needs as it explores options for replacing aging SMA training vessels; however, no decision has been made at this point regarding their replacement.

Environment and Compliance

Ship Disposal Program[15]

MARAD is the ship disposal agent for Federal Government-owned merchant-type vessels of 1,500 gross tons or greater and has custody of a fleet of non-retention ships. When ships are determined to be no longer of sufficient value to merit the cost of further preservation, MARAD arranges for their responsible disposal on a worst-first basis. Currently, MARAD has 18 obsolete vessels slated for eventual disposal. MARAD gives priority to expeditiously removing the vessels from the Suisan Bay Reserve Fleet (SBRF) as required by the April 2010 U.S. District Court Consent Decree. Of the 57 obsolete vessels once in the SBRF, two ships remain. The Consent Decree requires MARAD to remove all non-retention vessels by the end of FY 2017. The decline in domestic scrap steel prices makes it difficult for MARAD to sell vessels to recyclers and results in the need to pay to recycle ships. As a result, funds will be needed to pay to dry dock, hull clean, tow to the Gulf Coast, and dismantle these vessels if we are to meet the Consent Decree.

[15] See, 46 U.S.C. § 57102 for NDRF vessels and § 57101(c) for authority for other agencies to transfer vessels into the NDRF for disposal.

MARAD is also responsible for continuing the required protective storage activities for the inactive former Nuclear Ship SAVANNAH (NSS), including nuclear license compliance, radiological protection, ship maintenance and custodial care, and planning and preparation for decommissioning. The NSS decommissioning is to be completed by December 2031, which coincides with the current Nuclear Regulatory Commission license term.

Intermodal Development

Port Infrastructure Development[16]

Ports and the U.S. marine transportation system are critical to our economy. To better support our ports MARAD developed a port infrastructure development program called StrongPorts.[17] StrongPorts delivers tools and technical assistance to ports and integrate ports and maritime transportation into the larger U.S. surface transportation system. MARAD also oversees funding for port infrastructure projects provided through the Transportation Investment Generating Economic Recovery (TIGER) program authorized in the American Recovery and Reinvestment Act of 2009 (P.L. 111-5), and the Nationally Significant Freight and Highway Projects program which was authorized in section 1105 of the Fixing America's Surface Transportation Act (FAST Act), (P.L. 114-94). Since 2009, DOT awarded $578 million in TIGER funding for 48 port or marine highway projects in 27 states, and in FY 2016, awarded $115 million for five FASTLANE grant projects.

Short Sea Transportation Program (America's Marine Highways)[18]

Projects designated under the America's Marine Highways program make use of our Nation's vast network of waterways and coastlines to provide new export-based supply chain alternatives for our Nation's manufacturers and shippers. The mission of the program is to lead the development and expansion of services that move freight along our waterways and coastlines and to facilitate their integration into the U.S. surface transportation system. The program encourages partnerships with a variety of stakeholders including shippers and manufacturers, truckers, ports and terminals, ocean carriers, and domestic vessel operators to create new supply chain options that utilize our waterways. America's Marine Highway projects also allow for the optimization of

---

[16] See, 46 U.S.C. § 50302(c).

[17] StrongPorts is a collection of the Maritime Administration's programs and efforts aimed at improving ports. MARAD authority for intermodal development is based on several authorities including those you have noted. Additionally, MARAD has promotional authority for short sea shipping found generally in chapter 556 of title 46.

[18] See, 46 U.S.C. Chapter 556. The Consolidated Appropriations Ct of 2016 (P.L. 114-113), provided specific funding for the Short Sea Transportation Program.

equipment relocation and help to reduce wasteful movement of empty shipping containers. In FY 2016, MARAD awarded $4.85 million in Marine Highway Grants for six projects impacting nine states.

Shipbuilding and Financing

Maritime Guaranteed Loan Program (Title XI)[19]

MARAD's Title XI Program provides loan guarantees to enable successful applicants to secure long-term financing for shipyard modernization projects and for building vessels in U.S. shipyards. The loan guarantees provide applicants with long-term financing at favorable interest rates, while sustaining facilities for shipbuilding and ship repair within the United States. In FY 2017, the Title XI program issued a Letter of Commitment for the construction of two new LNG powered combination container roll-on/roll-off vessels.

Assistance for Small Shipyards and Maritime Communities[20]

The Small Shipyard Grant program provides funding to support capital improvements and employee training at small U.S. shipyards. Small shipyards play a significant role in our Nation's shipbuilding and repair activity. The grants support efficiency improvements and modernizations that allow U.S. shipyards to compete more effectively in the global market place. Congress has provided approximately $177 million between FY 2008 and FY 2016 for the Small Shipyard Grant program supporting 160 grants. In FY 2016, MARAD awarded $4.9 million in funding to 9 small shipyard projects.

Conclusion

We will continue to keep this Subcommittee apprised of the progress of our program activities and initiatives in these areas in the coming year.

I appreciate the Subcommittee's continuing support for maritime programs and I look forward to working with you on advancing maritime transportation in the United States. I will be happy to respond to any questions you and the members of the Subcommittee may have.

---

[19] See, Title XI of the Merchant Marine Act for 1936, as amended, codified at 46 U.S.C. Chapter 537.

[20] See, 46 U.S.C. § 54101.

HEARING ON THE "AUTHORIZATION OF COAST GUARD AND MARITIME TRANSPORTATION PROGRAMS," SUBCOMMITTEE ON COAST GUARD AND MARITIME TRANSPORTATION, TUESDAY, APRIL 4, 2017

2167 Rayburn House Office Building
Washington, DC
April 4, 2017
Submitted on behalf of Congressman Duncan Hunter (C4-50)

Questions for the Record to MARAD

The 2017 National Defense Authorization Act, NDAA (P.L. 114-840) was enacted into law on December 23, 2016. Section 3517, directed the Maritime Administrator, in consultations with the Coast Guard Merchant Marine Personnel Advisory Committee and the Committee on Marine Transportation System, to convene a working group to examine and assess the size of the pool of U.S. citizen mariners necessary to support the U.S. flag fleet in times of national emergency.

The working group was required to convene 120 days (April 26, 2017) after enactment:

Question 1. Has the working group been created?

Response: A charter for Maritime Workforce Working Group (MWWG) was created in early calendar year 2017 as a subcommittee of the U.S. Maritime Transportation System National Advisory Committee (MTSNAC), which operates under the Federal Advisory Committee Act. MARAD hosted the first meeting on March 10, 2017.

Question 2: How many people have been appointed?

Response: Twenty-six members, representing twenty-two stakeholder organizations, have been appointed to the MWWG. As specified in the 2017 NDAA the member organizations include MARAD, the U.S. Merchant Marine Academy, U.S. Coast Guard, the U.S. Navy and the U.S. Navy's Military Sealift Command, the State Maritime Academies, U.S.-flag fleet vessel owners operating in coastwise and international trades, and non-profit labor organizations representing licensed and unlicensed employees operating in the U.S.-flag fleet. Members of five organizations were also brought in as subject matter experts, including representatives from the U.S. Transportation Command, U.S. Army, and U.S. Department of Defense.

Question 3: Will the working group he able to produce the required report in one year, as directed?

Response: The MWWG is on target to complete its report within a year as mandated by Congress.

A Report from the Congressman Garamendi's Maritime Advisory Committee, Priorities of the Northern California Maritime Industry, Summary of Views Shared at a Roundtable Convened April 25, 2016, at the California State University Maritime Academy

The Maritime Advisory Committee, Roundtable Steering Committee

Captain Lynn Korwatch
Executive Director
Marine Exchange of the San Francisco Bay Region

Captain Bruce G. Clark
California Marine & Intermodal Transportation System
Advisory Council

Dave Nolan
Marine Personnel Manager
Patriot Services

Norman Fassler-Katz
California Marine & Intermodal Transportation System Advisory Council

Christian Yuhas
Marine Engineers Beneficial Association

Captain Tom Cullen
Administrator
California Office of Spill Prevention & Response

Jon Haveman
Principal
Marine Economic Consulting, LLC.

Steve Kreta
Vice President of Student Affairs
California State University Maritime Academy

Laura Podolsky
Policy Director
UC Davis
National Center for Sustainability Transportation

Jim Lites
Director
California Association of Port Authorities

Congressman Garamendi addressing his Maritime Advisory Committee.

Disclaimer

The roundtable was convened to sample the ideas and perspectives of maritime stakeholders in the Northern California region. While the Office of Congressman John Garamendi attempted to invite as many stakeholders as possible, our database is certainly not exhaustive, and not every invitee was able to attend. We invite any thoughts and comments not featured in this publication, and welcome additional participation at future Maritime Advisory Committee events. The views expressed in this document were shared generally at the roundtable meeting and should not be taken as the endorsed beliefs or recommendations of any single participant or organization.

Located in Vallejo, CA, the California State University Maritime Academy is the only degree-granting maritime academy on the West Coast.

The Training Ship *Golden Bear* (TS *Golden Bear*) serves as the training platform on which cadets can apply the technological and leadership skills they learned in the classroom and from their work assignments. Each summer, first and third-year cadets depart for two months during the annual Training Cruise.

Introduction

On April 25, 2016, Congressman John Garamendi (D-Fairfield, Davis, Yuba City, CA), Ranking Member of the Transportation and Infrastructure Subcommittee on the Coast Guard and Maritime Transportation, convened a roundtable discussion at the California State University Maritime Academy in Vallejo. More than 250 maritime stakeholders participated in this discussion, representing federal, state, and local agencies, maritime businesses, and labor unions. The roundtable had the following two goals:

1. **Broadly identifying key priorities of the many sub-sectors within Northern California's maritime industry:** blue/brown water steamship companies, waterfront industry (terminal operators, stevedoring companies, cranes & equipment operators), employer associations (PMSA), employee associations (ILWU & MEBA), ship servicing industry (Bar Pilots), and logistics & transportation industry (NVOCC, CA Trucking Association).

Eastern Shipbuilding Founder and CEO Brian D'Isernia describing the completed $100 million deep water oil service ship ready for delivery in Panama City, Florida -a Made in America Jones Act ship.

2. **Developing specialized working groups to discuss priority issues and goals in greater detail. Working groups included the following sectors/issue categories:** Private Sector, Public Sector, Education/Workforce Sector, and Environmental. Each working group identified and addressed their priority issues through the following 4-prong methodology:
   a) Description and explanation of the "state of issue" for each priority agreed upon by the sector working groups
   b) Significance of the priority issue (implications to private & public sector, etc.)
   c) Recommendations for Congressman John Garamendi regarding possible legislative solutions, best practices and/or action plans

d) Appendix of additional comments and suggestions from Advisory Committee Members

Working Group Issue Topics

A. Public Sector Funding & Initiatives	**(Working Group 1)**
1. 2016 Water Resources Development Act (WRDA)	
2. 2016 FAST Act Implementation in California ($23 Billion for CA)	
3. U.S. Port Security (Port Security Grant Program- FEMA)	
4. 2016 U.S. Coast Guard Re-Authorization	
5. M-580 Marine Highway Program	
6. Port & Terminal Congestion	
7. Federal agency response to the introduction of ultra-large containerships	
8. MARAD Title XI Federal Loan Guarantee Program	
B. Private Sector Economic Opportunities	**(Working Group 2)**
1. Jones Act Provisions & Proposals	
2. Revitalizing the U.S. Shipbuilding Industry	
3. Make it in America & Buy America Provisions	
4. West Coast Maritime Industry Implications of Panama Canal Expansion	
5. U.S. Coast Guard Waterway Management Policies & Forecast	
6. Port & Terminal Congestion	
7. Land Use & Urban Port Planning (Regional Private/Public Sector Partnerships)	
C. Regulations & Environmental Sustainability Goals	**(Working Group 3)**
1. CA Exec Order B-32-15 (CA Freight Mobility Plan & Sustainability Plan)	
2. Land-Use & Urban Port Planning (plans for building bulk commodity capacity)	
3. U.S. Endangered Species Act / WRRDA 2014 implications	
4. Terminal Emission Control Agency - Fuel Transition Goals & Plans	
5. California Air Quality Board & U.S. Environmental Protection Agency	
6. The need for National Emission Standards for the Maritime Industry	
7. U.S. Clean Water Act (Foreign Ship Ballast Water Issues)	
D. Education & Workforce	**(Working Group 4)**
1. Educating policymakers on the importance of the maritime industry to U.S. national security and economic security	
2. Building future U.S. Maritime Workforce via apprenticeships & WIOA (Workforce Innovation Opportunity Act) industry-driven curriculum	
3. Funding for Merchant Marine Training Ship Programs	
4. Coast Guard Licensing & Certification for Merchant Marine	
5. Shipbuilding Naval Architecture & Ship Design Programs	
6. Implementation/Discrepancies in Jones Act Provisions	

Priority Issue Survey

Issue Category	Description Of Priority	Total
Private Sector	Jones Act Provisions & Proposals	27
Private Sector	Port & Terminal Congestion (Private Sector Implications of Pier Pass & Oakland Pier Pass Proposal)	15
Private Sector	Revitalizing the US Shipbuilding Industry	14
Education/Workforce	Educating Policymakers on the importance of Maritime Industry to the local, regional, & national economy	13
Private Sector	Make it in America & Buy America Provisions	10
Public Sector	M-580 Marine Highway Program	8
Regulations & Env.	CA Executive Order B-32-15	7
Regulations & Env.	CA Air Quality & US Environment Protection Agency	6
Public Sector	US Port Security (Port Security Grant Program - FEMA)	6
Public Sector	Federal Agency Response to the introduction of Ultra-Large Containerships	5
Public Sector	2016 Water Resources Development Act (WRDA)	5
Private Sector	Land Use & Urban Port Planning (Regional Private/Public Sector Partnership)	5
Education/Workforce	Building future US Maritime Workforce (Ex: via apprenticeship program, etc.)	5
Education/Workforce	Coast Guard Licensing & Certification for Merchant Marine	5
Regulations & Env.	The Need for NTL Emission Standards for the Maritime Industry	4
Private Sector	West Coast Maritime Industry Implications of Panama Canal	4
Public Sector	MARAD Title XI Federal Loan Guarantee Program	4
Public Sector	2016 Coast Guard Re-Authorization	3
Public Sector	2016 Fast Act Implementation inCA ($23 Billion for CA)	3
Education/Workforce	Implementation/Discrepancies in Jones Act Provisions	3
Regulations & Env.	Terminal Emission Control Agency Fuel Transition Goals & Plans	2
Regulations & Env.	US Clean Water Act	2
Private Sector	US Coast Guard Waterway Management Policies & Forecast (Coast Guard's Role in Ice-Breaking Efforts)	2
Education/Workforce	Shipbuilding Naval Architecture & Ship Design Programs	2
Education/Workforce	Funding for Merchant Marine Training Ship Programs	2
Regulations & Env.	Land Use & Urban Port Planning	1
Regulations & Env.	US Endangered Species Act/ WRRDA 2014 Implications	0

Congressman Garamendi engaging with his Education & Workforce working group during the Maritime Advisory Committee Roundtable.

Congressman Garamendi providing remarks during a Change of Command Ceremony at Coast Guard Island Alameda.

Congressman Garamendi boarding a Coast Guard patrol boat for a Mississippi River tour.

Summary of Recommendations and Priorities

Public Sector Recommendations and Priorities

1. Congress should adjust the definition of "Donor Ports" within WRDA, to reflect a $5 million contribution rather than the current $15 million contribution to the Harbor Maintenance Trust Fund.
2. Congress should continue supporting legislation such as the FAST Act that emphasizes programmatic funding for multi-modal freight transportation, such as rail, highways, marine-highways, and ports, as an interconnected and interdependent system of systems, and which provides a strategic framework for multi-modal transportation systems both at a national and state level.
3. Congress should provide additional dedicated funding for "port security & maintenance" through additional discretionary funding programs and grant funding dedicated to port authorities and port stakeholders, where contributory equity is a key factor in assuring a fair return and distribution of tax and fee payer revenues dedicated for these purposes.
4. National and local policies/practices at our ports should include "coordination" of infrastructure investments and implement practices such as: 1) Pre-Arrival Scheduling, and 2) coordinated infrastructure development and deployment (federal, state, local, public and private), both landside and waterside to address port congestion issues. As a result of this recommendation, Congressman Garamendi supported and monitored the activities of the Port Performance Freight Statistics Working Group authorized under the FAST Act, as well as the work of the FMC's Supply Chain Innovation Initiative.
5. Both national and statewide leadership should revisit the deployment and implementation of the M-580 program and similar strategies for coastwise short sea and inland river transportation as a means to address: 1) Port Congestion, 2) Compliance with 2016 California Freight Sustainability Goals and associated environmental target goals, and 3) economic opportunities to strengthen the economic viability of the maritime industry.

Private Sector Recommendations and Priorities

1. Congress should continue advancing the Jones Act, while considering revisions that can strengthen its original intent of promoting the U.S. maritime industry and protecting national security interests. Congress should consider: 1) possible revisions in the "build and charter" program; 2) revising the construction differential subsidy and operating differential subsidy; and 3) including

provisions requiring American new ship builds and American existing ship repairs.

2. Congress should incentivize U.S. private sector leaders to construct additional U.S.-built vessels through Title XI tax incentives and implementation of additional policies, such as H.R. 5270 (113th Congress), to ensure that crude oil and LNG exports are used to expand our U.S.-flag fleet, U.S. mariner pool, and U.S. shipbuilding capabilities. Congress should incentivize the private sector to build U.S. ships utilized in international trade, and should prioritize enforcement of cargo preference rules. In keeping with this recommendation, Congressman Garamendi introduced H.R. 1240, the Energizing American Maritime Act, requiring that 30% of exported crude oil and LNG travel on U.S.-flag ship by 2025.

The USCCC Polar Star Icebreaker during dry dock repairs at Mare Island, CA. The Polar Star is a key Coast Guard asset in protecting American interests and ensuring safe and reliable maritime commerce.

Environmental Sustainability Recommendations and Priorities

1. Congress and/or the state legislature should provide additional funding for tools and resources for the maritime industry to address the many environmental sustainability goals that have become increasingly burdensome and "unviable" for the industry's capabilities. Without such funding and resources, it will not only come at the cost of the economic competitiveness of the U.S. maritime industry but will disrupt long-term compliance of these sustainability goals.
2. Congress should propose a nationwide freight sustainability standard that works in conjunction with state standards and is mindful of the national maritime industry's needs and goals. If sustainability goals are ambitious, additional funding should be provided to ensure that these goals can be achieved.

Congressman Garamendi touring the Eastern Shipbuilding Naval Shipyards.

3. Congress and state policymakers should consider the following principles when developing new environmental policies impacting the maritime industry: 1) Establish uniform standards across all states that accurately reflect existing Best Available Technology, and establish goals that are reasonably attainable. 2) For regulations reliant on the adoption of new technologies, consider the viability and availability of respective technologies to the industry and adequate funding support for research & development. 3) For regulations that are performance-based, provide incentives and rewards for good actors.

Education and Workforce Recommendations and Priorities

1. Congress, state policymakers, and maritime industry stakeholders should coordinate efforts to advance public discourse on the priorities and needs of the maritime industry by: 1) holding public hearings and additional roundtables that are work-focused and outcome-driven; 2) planning field trips for students to ports; 3) hosting larger lobbying efforts; 4) hosting maritime industry & policymaker lunch meetings.
2. Congress should work with the Maritime Administration and the U.S. Coast Guard to ease licensing requirements and programs for veterans entering the maritime workforce, noting that many veterans are currently not receiving credit for training completed during their military service. Many veterans already possess expert knowledge and training, but face delay in entering the industry following release from active duty due to redundant licensing requirements and

fees that delay or completely disrupt the transition into the maritime industry at a time when the maritime workforce faces critical projected shortfalls. In keeping with this recommendation, Coast Guard and Maritime Transportation Chairman Duncan Hunter and Ranking Member John Garamendi convened two separate listening sessions with Federal agencies and maritime stakeholders on the issue in 2016, and were successful in including "Military to Mariner" language in the 2017 National Defense Authorization Act.

3. Congress and U.S. private sector interests should work together to develop a new era of life-saving equipment and updated requirements for U.S. mariners and U.S. vessels. With climate change, significant weather events are growing more severe every year, and new equipment and training will be critical in preventing further loss of American lives and shipping infrastructure, along with the associated economic and social impacts.

Public Sector Priorities

Public Sector Working Group Facilitator:
Captain Tom Cullen Administrator
CA Office of Spill Prevention & Response

Priority Issue #1 -Water Resources Development Act (WRDA) 2016

State of Issue

The Harbor Maintenance Trust Fund (HMTF) provides funding for port dredging and maintenance paid for by a 0.125 percent user fee on the value of cargo shipped through a port. A port is considered a "donor port" if it collects at least $15 million in HMTF dollars annually, received less than 25 percent of collected HMTF dollars in the last five years, and is located in a state where over 2 million twenty-foot equivalent units were loaded or unloaded in FY 2012. Donor ports qualify for assistance from a $50 million set-aside that can be used for berth improvements and dredged materials disposal and reuse, environmental reviews, and payments to importers. California, whose northern ports face a dire need for additional dredging funding, is a major donor state that, as *of* 2015, received only 15 percent of the amount paid into the HMTF by shippers for the use of its harbors. For example, shippers at the Ports of Los Angeles and Long Beach paid on average $250 million each year into the HMTF, but those ports received an average of $2 million in return.

Significance

California has many needs to maintain our ports and canals. Dredging operations, utility improvements, and infrastructure upgrades are just a few of the many costs our ports face to remain competitive.

Container Ships in the Port of Oakland. Photo: Steve Boland, http://calurbanist.com/.

Recommendations

Congress should put pressure on Appropriators to allocate additional funds for dredging, and should adjust the definition of "Donor Ports" within WRDA to reflect a $5 million contribution rather than the current $15 million contribution to the Harbor Maintenance Trust Fund. The Sector Group particularly recognizes the efforts of the Port of San Diego and the Port of Hueneme in this regard, as shown in the following example:

a) San Diego contributes an average of between $6 million and $8 million annually.
b) The current $15 million contribution requirement does not qualify the Port of San Diego for "expanded use" of HMTF funds, which include berth dredging and maintenance projects and disposal of contaminated soil.
c) The Port of Hueneme also supports the adjustment of $5 million.
d) The Executive Directors of the Ports of Los Angeles and Long Beach are not opposed. Congressional delegations representing San Diego and Ventura have expressed support.

Note: With Congressman Garamendi's support, the Water Infrastructure Improvements for the Nation Act (WIIN), the 2016 version of WRDA, was passed and signed into law in December of 2016. It included a provision to expand the donor port program to include medium-size ports that collect between $5 million and $1 5 million for the HMTF annually, received less than 25 percent of those funds back during the previous five years, and are located in a state where more than 2 million shipping containers were loaded or unloaded in 2012. This is a win for the Port of San Diego, the Port of Hueneme, and all of California.

Priority Issue #2- 2016 Fast Act Implementation

State of Issue

In December of 2015, a five-year surface transportation bill was signed into law after nearly a decade of short-term extensions. The $305 billion plan, called the Fixing America's Surface Transportation (FASD Act, highlighted freight infrastructure as a top national priority with the first ever freight title and over $10 billion directed for freight programs.

Significance

The FAST Act marks the first time the federal government proactively recognized the need to fund multi-modal port systems. Money for freight is incredibly important, supporting a larger statewide freight transportation system that serves America's businesses and consumers. Freight grants made available through the FAST Act, like the FASTLANE grant program, will be important in alleviating the high demand for TIGER grants.

Recommendations

Congress should continue supporting legislation such as the FAST Act that emphasizes programmatic funding for multi-modal freight transportation (rail, highways, marine-highways, etc.) and provides a strategic framework for multi-modal transportation systems both on a national and state level. Such legislation must recognize the important role ports play in freight movement, and stakeholders should strive to get ports projects listed in the Transportation Improvement Program (developed by Metropolitan Planning Organizations) and Statewide Transportation Improvement Program (developed by California).

Priority Issue #3 - Port Security Grant Program

State of Issue

The Port Security Grant Program (PSGP) supports the building, sustainment, and delivery of core capabilities essential to national security, directly supporting maritime transportation infrastructure security activities. Over a five-year period from 2006 to 2010, a total of nearly $1.7 billion in PSGP funds were allocated, averaging $340 million annually. But the program has seen significant cuts, and in FY2016 was funded at just $100 million for FY2016. Further, these funds are now allocated on a purely competitive basis by a central panel in Washington, DC, rather than based on the needs assessments of regional partners most familiar with local priorities and requirements.

Significance

Ports often lack the sophisticated lobbying efforts of other modes of transportation and are often unable to compete for scarce grant dollars. This grant program helps state, local, and private sector maritime partners improve port-wide risk management, maritime domain awareness, training, and recovery and resilience capabilities.

Recommendation

Congress should provide additional dedicated funding for port security and maintenance, including through additional discretionary funding programs and grant funding dedicated to port authorities and port stakeholders. Congress should:

a) create a dedicated funding source for Cyber Security and Radiological/Nuclear Threats and Protections (RADNU) that focuses specifically on these threats
b) keep the Port Security Grant Program as a unique budget line item and resist moves to consolidate it into other grant initiatives;
c) ensure tax dollars collected in California that fund the PSGP fairly and equitably benefit California Ports and Harbors, and that prioritized projects are selected using regional input as the primary consideration;
d) streamline and revise the PSGP to extend the grant expenditure timeline (currently limited to a two year expenditure cycle) to improve obligation and expenditure rates for PGSP funds (some participants expressed views that local project approvals and federal and state environmental compliance requirements can erode the actual project performance schedule to an unattainable level, particularly for civil improvement and protection projects); and
e) expand the approved use of funds to include maintenance and replacement of critical port security equipment and infrastructure with a limited effective life cycle, whether previously purchased with PSGP funds or not.

Note: In keeping with this recommendation, Congress convened two hearings in 2016 to gather additional background on port security threat assessment and security.

Priority Issue #4 - Port & Terminal Congestion

State of Issue

Outdated infrastructure, inefficient processes, labor disruptions, and changing market trends (including the introduction of ultra-large carrier vessels and the use of megaships) are among the many factors that contribute to port and terminal congestion. Federal assistance is needed to address channel depth, landside infrastructure, intermodal connectors and other factors that can contribute to the smooth functioning of ports. Concerns include scheduling costs, overtime labor, labor management issues, and insufficient access to ports and transactions.

Port of Oakland aerial view. Photo: Daniel Parks, https://www.flickr.com/photos/parksdh/.

The Committee specifically identified concerns about Pier Pass, a program unique to the Port of LA and Port of Long Beach. They pointed out that Pier Pass Scheduling covers only two shifts during a five-day period and allows drayage companies and shippers to pay a "fee" for preferential assignment of a "pick-up" slot, when not all carriers and shippers have the resources and flexibility to do this. The Committee also noted that the warehousing and inland port staging areas often do not match extended gate hours at the port facilities, so that even if the port gates remain open for three shifts, drayage companies have no place to go with their loads. These elements limit the attainment of maximum efficiencies, resulting in port congestion and the accompanying economic and environmental impacts.

Significance

Businesses, workers, and consumers across the U.S.- and the world- rely on the smooth and efficient flow of goods through our ports and terminals to remain competitive. Port and terminal congestion threaten the efficiencies of supply chains crucial to our global economy, and challenge implementation and attainment of rigorous "zero emissions" environmental goals.

Recommendations

Noting that a proposed "increased fee" for usage negatively impacts the trucking industry, the Committee determined that a comprehensive approach is needed to address port congestion. Both national and local port decision makers should pursue "coordination" of infrastructure investments and implement practices such as: 1) pre-arrival scheduling; 2) coordinated infrastructure deployment both landside and waterside to addressing port congestion issues; and 3) avoidance of limited gate hours, including the avoidance of gate closures on Sundays.

Port of Oakland cranes. Photo: Steve Boland, http://calurbanist.com/.

Containers being moved in preparation for loading at the Port of Oakland.
Photo: Department of Energy.

Priority Issue #5- the M-580 Marine Highway Program

State of Issue

The M-580 Marine Highway Program was discussed at Congressman Garamendi's 2013 Maritime Roundtable and supported by the Congressman. Unfortunately, a series of issues led to the failure of the California Test Project on the Sacramento River between the Ports of Oakland and Stockton, including a lack of time and resources for full implementation. For instance, Virginia's Marine Highway Program was given three years of funding for private-market value sustainability and acceptance, whereas the M-580 only received 14 months of funding. Some Committee members also offered the idea that the selection of this relatively short inland river route may not have been the best initial test of this concept in California.

Congressman Garamendi speaking on the House Floor about the importance of building a US. polar icebreaking fleet.

Significance

The Public Sector Group identified multiple reasons to secure additional funding to revive this program, including:

a) Marine highways are proven tools that can address port congestion, roadway traffic, and general wear-and-tear on inland highway infrastructure, while promoting improvements and increased service by rail to currently underserved areas.
b) With the State of California mandating stricter Zero Emission sustainability goals, the marine highway can be a valuable and effective alternative.
c) Because the infrastructure for this program is already in place, subsidies are only required to assist with temporary cargo movement.
d) There is a strong business case for the M-580 Program when considering cargo perishability issues for conventional cargo transportation.

Recommendations

Both national and statewide leadership should revisit the deployment and implementation of the M-580 program as a means to address port congestion, compliance with California's Freight Sustainability Goals, and opportunities to strengthen the economic viability of the maritime industry by development and construction of U.S.-made ships crewed by U.S. mariners.

From left: Senator Chris Murphy, Rep. Ted Deutch, former Rep. Donna Edwards, Rep. John Garamendi, and Senator Brian Schatz aboard the USS Mustin.

Congressman Garamendi touring the USCGC Polar Star; Icebreaker during dry dock repairs at Mare Island, CA.

Private Sector Priorities

Private Sector Working Group Facilitator:
Jon Haveman
Principal
Marin Economic Consulting, LLC.

Priority Issue #1 -Jones Act Provisions

State of Issue

The Jones Act is vital to the U.S. maritime industry. The Sector Group focused on the importance of protecting the Jones Act, as well as revisions that adhere to its original goals and strengthen many of the sub-sectors of the U.S. maritime industry, including the revitalization of the U.S. shipbuilding industry.

Significance

The Jones Act not only has major economic implications for the maritime industry, but also significant and serious implications for the national security of our nation. The current composition of the Jones Act fleet is both insufficient for the long-term viability of the maritime industry and poses a national security concern that must be addressed.

Top and Bottom: Congressman Garamendi touring Bollinger Shipyard in Louisiana, where 22 fast Response Cutters have been built for the Coast Guard so for.

Recommendation

Congress must continue advancing the Jones Act, while considering revisions that can strengthen its original intent of promoting the U.S. maritime industry and protecting national security interests, including:

a) Possible revisions in the "build and charter" program.
b) Possible revisions to the construction differential subsidy and operating differential subsidy.
c) Revisions to require that all repair work on U.S.-flag Jones Act vessels-with an exception for emergency repairs-be done in the United States in U.S. shipyards with U.S. labor and materials.

Note: In keeping with this recommendation, Congressman Garamendi and others were successful in fending off efforts in 2016 to weaken the Jones Act.

Priority Issue #2 - Revitalizing the U.S. Shipbuilding Industry

State of Issue

In 1979, the United States had over 576 U.S.-flagged vessels operating in the foreign trade. Today, that number has plunged to under 90 vessels. This slow decline in U.S. shipbuilding and ship operating capacity poses great concerns for America's long-term economic viability and national security.

Significance

There are only nine remaining active blue-water (large vessel) shipyards capable of building vessels for foreign trade. Moreover, as noted the current build (construction schedule) of "Jones Act Fleets" will not be enough to fulfill all economic needs and/or U.S. national security interests. Without enough Jones Act vessels, many ship owners are left to reflag foreign vessels, a practice that has short-term functional or economic value for civilian cargo but presents great national security risks and concerns for military cargo lift capacity over the long term.

Recommendations

Congress should incentivize U.S. private sector leaders to construct additional U.S.-built vessels through Title XI tax incentives and implementation of additional policies such as H.R. 5270 to ensure Crude Oil and LNG oceangoing exports utilize U.S.-built, -crewed, and -flagged ships. Congress should provide incentives for private sector construction of U.S. ships utilized in international trade, and should prioritize enforcement of cargo preference rules. Congress should pursue and promote:

a) policies that encourage or require the use of U.S. ships for international trade;
b) policies and practices that include dual military vessel usages, including incentivizing the development and use of coastal container and bulk carrier vessels which, under Jones Act provisions, can be made available to the Department of Defense in times of war, and exploring other military and civilian joint usage ideas;
c) policies that provide better shipbuilding opportunities;
d) policies that increase U.S. qualified mariner base;
e) tax incentives that encourage the construction of U.S. ships; and
f) export trade policies that protect and promote the U.S. maritime industry, including requirements that set volumes of export cargoes must travel on U.S.-flag vessels.

Note: In keeping with this recommendation, Congressman Garamendi introduced H.R. 6544 and H.R. 6545 at the end of the 114th Congress, as well as H.R. 1240 in the 115th Congress. He also supported colleagues in the passage of a 2017 National Defense Authorization Act provision authorizing the design and construction of a new multi-mission maritime training vessel for the U.S. Merchant Mari ne Academy and state maritime academies.

Congressman Garamendi delivering the keynote address during the 2016 SF Fleet week Senior Leaders Seminar.

Sustainability and Regulations Priorities

Sustainability Working Group Facilitator:
Laura Podolsky
Policy Director, University of California, Davis
National Center for Sustainability Transportation

Priority Issue #1 - Funding for Near Zero Emissions Equipment

State of Issue

California has the strictest environmental regulations for the maritime industry in the United States. The general consensus of the working group was that these are good and sound policies in attaining a balance between economic, commercial, and environmental goals. But because other states have been slow or reluctant to adopt similar regulations, the stage has been set for uneven and unequitable competition.

Significance

Federal preemption by establishing a set of national standards could help expand uniform application of environmental regulations and result in the following things:

a) California would not bear the entire cost of bringing new and more efficient engines and water treatment systems to market, and maritime commercial operators would not be required to purchase these systems to comply only with the requirements of a single state.
b) Cargo companies would not be tempted to forgo California ports for ports in other states because of the cost of compliance with California's environmental regulations. If more cargo is diverted to other ports, California is not fixing the problem of reducing greenhouse gases and is instead exacerbating the problem by transferring these issues to other regions and states. Greenhouse gases would be reduced on a nationwide level by the adoption of uniform national standards focused on attainable and realistic goals and using a "Best Available Technology" approach.
c) What should come first: more efficient products, or regulations that require more efficient products? Without stricter regulations, there is no incentive to create more efficient products, but regulations must come with funding for research and development, and incentives for bringing technologies to market. Regulations on vessel incidental discharges, described in a subsequent section, are a good example of the fragility of this balance.

Recommendations

Congress and the state legislature should provide additional funding for tools and resources for the maritime industry to address the many environmental sustainability goals that have becoming increasingly burdensome and "unviable" for the industry's capabilities. Without such funding and resources, it will not only come at the cost of the economic competitiveness of the U.S. maritime industry but will disrupt long-term compliance of these sustainability goals.

Priority Issue #2 - California's Sustainable Freight Action Plan

State of Issue

Governor Jerry Brown's executive order B-32-15 directed the California Air Resources Board (CARB), the California Department of Transportation (Caltrans), the California Energy Commission (CEC), and the Governor's Office of Business and Economic Development "to develop an integrated action plan by July 2016 that establishes clear targets to improve freight efficiency, transition to zero-emission vehicles (ZEVs), and increase competitiveness of California's freight system." The integrated plan must be informed by existing state agency plans, most notably the California Freight Mobility Plan (produced by Caltrans), Sustainable Freight Pathways to Zero and Near-Zero Emissions (produced by CARB), and the Integrated Energy Policy report (produced by CEC).

Significance

Current projections show that reconciliation of these plans will come at considerable cost to the California inter-modal transportation system. There are still many questions that have yet to be answered, including:

a) How will the trucking and the freight industry at large transition to ZEVs?
b) Will the action plan be prescriptive?
c) Although many California agencies are working together, how will the state's efforts fit into the federal context? If California efforts do not sync with the federal dynamics, the Sector Group is concerned that the state may become less economically viable.
d) How will regulations affect the cost of moving cargo?
e) What's next after the statewide plan is adopted?

Recommendations

Congress should propose a nationwide freight sustainability standard that works in conjunction with the state standards, and is mindful of the national maritime industry's needs and goals. If sustainability goals are ambitious, additional funding should be provided to ensure that these goals can be achieved.

Priority Issue #3- Reconciling Technological Capacity of Industry with National & State Environmental Regulations and Sustainability Goals

State of Issue

With new environmental regulations and sustainability goals being adopted both by the state and national government on an on-going basis, maritime industry sectors face challenges in the ambitious nature of many of these reforms, as well as inconsistencies between state and federal standards. Leaders in every sector of the maritime industry understand the importance of adopting sustainability goals, but request that policymakers consider strategies to assist in achieving sustainability goals while remaining economically competitive. One key issue is ensuring that regulations and goals are being established in consideration of the viable technology available to industry. Policymakers must also ensure that available technology is viable for large-scale commercialization and implementation to meet the demands of new regulations on a statewide and national scale.

Significance

On July 17, 2015, the Governor's Executive Order B-32-15 to coordinate the California Sustainability Freight Action Plan with the goals of improving freight efficiency, transitioning to zero-emission technologies, and increasing economic competitiveness in the state. While a positive step forward, the Sector Group identified concerns in the maritime industry of possible opportunity losses related to this order. After the order was announced, the California State University Maritime Academy Golden Bear Research Center began identifying segments of the freight sector that would face challenges in adopting the mandated new technologies- an indication that regulations and sustainability goals were set without clear evidence that they were reachable. The Sector Group notes that industry often takes a triple-hit when environmental regulations and standards change: once in up-front R&D costs, again for commercialization and implementation of the required technology, and finally in competitive impacts if the associated costs are not spread universally across the national marketplace.

Recommendations

Federal and state policymakers should consider the following principles when developing new environmental policies impacting the maritime industry:

a) Strive for uniform standards across all states.
b) For regulations reliant on the adoption of new technologies, consider the viability and availability of respective technologies to the industry, provide adequate funding support for R&D, and incentive partnerships for implementation.
c) For regulations that are performance-based, provide incentives and rewards for good actors.

Note: The International Maritime Organization's International Convention for the Control and Management of Ships' Ballast Water and Sediments (BWM) will enter into force on September 8, 2017. Vessels operating under the flag of a signatory country must comply with those regulations, as well as with applicable port regulations issued by the

U.S. Coast Guard, the U.S. Environmental Protection Agency, and any state-specific regulations. In December of 2016, the U.S. Coast Guard approved Optimarin AS' UV-based treatment system for compliance with its regulations, with expectations of two additional approved systems shortly. While this is a positive step forward in clarifying this complex compliance environment, maritime stakeholders remain concerned about the evolving regulatory program and point out the need for a stable, uniform standard.

Education and Workforce Priorities

Education & Workforce Working Group Facilitator:
Steve Kreta
Vice President of Student Affairs
California Maritime Academy

Priority Issue #1 - Raise Awareness of Policymakers and the General Public on the Critical Importance of a Strong U.S. Maritime Industry to Our Economy and National Security

State of Issue

The U.S. maritime industry contributes more than $649 billion annually to national GOP, sustaining more than 13 million jobs. Virtually 90 percent of all cargo comes through our ports via the maritime industry, touching the lives of every American on a daily basis. Yet most policymakers and members of the general public are largely unaware of these vital contributions, the needs of the industry, and the various supply chains vital to its success (including trucking and rail).

Significance

Noting that the U.S. Maritime Industry is largely driven by legislative policies (including infrastructure funding, subsidies, and regulations), it is vital for U.S. policymakers and for the public at large to learn more about the value and impact the maritime industry has on our country's economic and national security. The economic footprint of this industry spans many sectors and is deeply tied to the strength of our local workforce. Expanding our U.S.-flag fleet and employing more workers, both on land and at sea, means more mariner and manufacturing jobs strengthening the entire U.S. economy. But due to a pattern of depreciating competitive value on a global scale, the percentage of U.S.-flagged and operated ships continues to decrease, to the point that our maritime industry now falls far short of its potential. As a result, America falls short of its potential as well.

Recommendations

Efforts should be coordinated amongst Members of Congress and state policymakers to elevate public discourse on the priorities and needs of the maritime industry, including: 1) holding public hearings, 2) additional roundtables, 3) planning field trips for students to ports, 4) hosting larger lobbying efforts, 5) maritime industry & policymaker lunch meetings, etc.

Congressman Garamendi Learning about the mechanics of the USCGG Polar Star Icebreaker.

California's community colleges, Like Solano Community College, can teach important engineering and physics concepts for those aspiring to enter the maritime economy.

Suggestions

a) Hold More Public Events and Hearings - Congressional meetings in the district are great ways to bring everyone together. Encourage policy makers to hold their own district roundtables. This year's Roundtable format is a great model and format for future discourse.
b) Lobbying and Marketing - Get policymakers physically on the ships, possibly to include a lunch date with the labor unions. It makes a big difference when they're on the ship and see the various operations for themselves.

c) Student School Involvement - Encourage more field trips to U.S. ports, train stations, railways, etc. Get the word out that there are good paying jobs within the maritime industry.

Priority Issue #2 - Building the Maritime Workforce

State of Issue

The maritime industry lacks a sufficient workforce and work opportunities. There is an insufficient number of ships and mariners to maintain the competitiveness and economic needs of the maritime industry, notably described by a recent MARAD report indicating that thousands of additional mariners will be needed in the near future to meet sealift requirements. To address these needs, we must build a pipeline by continually recruiting and training future mariners.

Significance of Issue

Currently there are impediments in certain industry and education sectors which act to decrease the speed and efficiency of rebuilding this workforce pipeline, specifically the transition of sailors and mariners from the military into the commercial maritime industry.

Recommendations

Congress should ease licensing requirements and programs for veterans entering the maritime workforce, noting that many veterans are currently not receiving credit for training completed during their military service. Many veterans already have the knowledge and training, but face delay in entering the industry due to redundant licensing requirements and/or fees. This can delay or completely disrupt the transition into the maritime industry. Congress should also focus on retention of those in the mariner industry, easing the many challenges mariners face in continuing an at-sea lifestyle.

Suggestions

Congress should do the following:

a) provide more flexibility for students coming from different backgrounds to help meet STCW guidelines;
b) require the Coast Guard and Navy to streamline the licensing process involved in the military-to-mariner transition to provide a career track option supporting this goal;
c) continue encouraging industry to credit military transition to workforce;
d) increase availability and affordability of mandated Advanced Training, noting the many new courses mandated by Coast Guard and international regulations

(courses should be made available online or outside of traditional working hours so participants are not forced to take time off);

e) eliminate federal income tax for mariners to encourage them to live at sea; continue advanced training to assist with mariner retention; and
f) better marketing to future mariners, highlighting humanitarian efforts, and a focus on getting youth connected through social media marketing.

Priority Issue #3- Naval Architecture (Ship Building & Design)

State of Issue

The Sector Group notes that there is not enough life-saving equipment on ships. Recent incidents with American lives lost at sea, like the sinking of the S.S. El Faro on October 1, 2015, should encourage policymakers and the private sector to investigate ways to prevent such tragedies from occurring in the future.

Significance

As climate change continues to increase the frequency and severity of storms, advances in life-saving technologies are needed to keep pace. Improving such technologies will decrease loss of life associated with maritime professions, alleviate instances of pain and suffering for family members, and improve the attractiveness of the maritime industry at a time when the industry is aging out and must place special emphasis on attracting young entrants.

Recommendations

Life rafts are not the answer. As a world super-power, we must strive to build more effective and cost-efficient life-saving vessels and equipment for both commercial vessels and cruise ships. Congress and U.S. private sector interests should work together to develop a new era of life-saving equipment for U.S. mariners, leveraging the R&D of the International Maritime Organization and other worldwide maritime organizations when possible.

Congressman Garamendi touring Coast Guard Sector New Orleans.

APPENDIX ONE: FOCUS COMMENTS

For Clarification of the 2016 Garamendi Maritime Roundtable Report Provided by Capt Bruce G. Clark, USCG (RET)- Chair, the California Marine and Intermodal Transportation Advisory Council

Public Sector Recommendation #1 "Adjust the Definition of "Donor Ports" within WRDA"

While it is advisable to evaluate the level of, and means testing for, financial support paid by each Port into the HMT fund, the more critical issue is to rebalance the payments from the fund to "spend down" the available funding as intended by Congress for dredging and harbor maintenance projects. Every year, funds in the Harbor Maintenance Tax Fund (HMT Fund) go unspent while available approved projects remain undone or incomplete. Donor states contributing the most funding should receive prioritized return on investment commensurate with the taxes paid, and the HMT fund should be paid down every fiscal cycle by clear direction from Congress and regulatory requirement. Each maritime state (coastal, Great Lakes and inland and western rivers) should also be required to contribute to the HMT goals and objectives through annual assessment contribution of matching funds at some level established by Congress. This action assures an equality of stakeholder investment (federal, state, and local) and recognition of the maritime transportation system (MTS) as a regional, state, and national asset- and not simply a national (federal) responsibility.

Public Sector Recommendation #4 "Coordination" of Infrastructure Investments"

Federal, state and local agencies and authorities must work together to manage public revenues targeted for MTS-related infrastructure investments. Funding in the past has been delayed, mired in competitive confusion, and "siloed," dependent upon agency focus and transportation mode bias. Clearinghouse activities to jointly prioritize and approve projects in an expeditious manner remains a critical challenge to assure maximum comprehensive benefit is obtained from the expenditure of limited public funds. A clear and effective approach to these complex issues can only be accomplished where the critical needs are evaluated across all modes of the MTS- not simply with a focus on a single mode requirement. As no transportation mode operates in isolation from the others, prioritization of critical projects must consider effects and benefits to all modes- and this requires a concerted and sustained, organized and well-coordinated effort across a full spectrum of government, agency, and MTS stakeholders. U.S. DOT MARAD should be fully empowered and funded to lead and manage this national effort, leveraging and utilizing the existing Maritime Transportation System National Advisory

Council (MTSNAC) and the federal TIGER Grant process, but in direct consultation and coordination with state equivalents such as the California Marine and lntermodal Transportation Advisory Council (CALMITSAC - a state level MTSNAC counterpart). CaiSTA and CFAC should lead the state prioritization effort, and the state, regional, and key county municipalities (where there are direct benefits and port entities) must all be willing to contribute infrastructure enhancement funding for selected critical projects.

Congressman Garamendi meeting with Maritime representatives in his Washington, D.C. office.

Public Sector Recommendation #5 "Revisit the Deployment and Implementation of the M-580 Program"

Complementing federal and local port initiatives, the State of California must join federal and local government and stakeholders to take a leadership role in supporting the development and trial of the M-580 concept over a sufficient timeline to assure a legitimate proof of concept. The State provided no funding supporting for the M-580 initiative and was only a marginal voice in attempting to reconcile differences between port, vessel operators, and labor considerations. With the advent of California's new environmental goals and objectives that further restrict air and water emissions - specifically a renewed focus on reducing congestion and pollutant concerns associated with highway transportation from major urbanized port regions - it should convene a regional working group to evaluate lessons learned from the initial effort of the M-580 initiative, and should analyze success stories available from test programs in other regions of the country to revitalize and relaunch the effort. Because the majority funding would likely again stem from federal sources, the U.S. Maritime Administration should again lead this process, supported and informed by an active and engaged coalition of state agencies (lead by CaiSTA and leveraging the California Freight Advisory Council) with the full partnership of labor and the trucking industry. Human nature often drives us all towards personal interest decision making, and change often means a shift in economic beneficiaries and job creation, but the attainment of mutually desirable long-term social and environmental goals and objectives requires a willingness to review the efficacy of practices and procedures to seek better solutions to urban goods movement challenges.

Further, a realistic exploration of an M-5 type offshore hub-and-spoke marine reception and distribution centers, such as the Portunus Project envisioned by researchers at the Lawrence Livermore National Laboratory (LLNL), might well address a combination of environmental, operational, and national security challenges that would completely rework traditional/and-based port operations and distribution mechanics, while promoting the development of smaller, cleaner coastwise feeder ships (U.S.-Flag/Jones Act Crewed) and possible regional feeder ports (Humboldt Bay, Monterey Bay FT Ord/Seaside, Pismo Beach/San Luis Obispo).

Where once California and other coastal states had a robust coastwise maritime industry along its long coastline, most of these once vibrant and active ports have declined or vanished with the advancement of rail and highway services. These smaller "ports" are now currently underdeveloped, underserviced, and economically vulnerable, but would significantly benefit by increased land-based facility and goods management jobs likely to be union-organized and compensated at sustainable, middle class "living wage" standards.

We expect these activities to change the transportation ecosystem while maintaining jobs tied to highway transportation, and to possibly revitalize short line rail service in these regions. Both elements would directly reduce congestion in urban areas notorious for heavy truck traffic and add jobs in historically-impacted areas. Long haul firms would still be contracted to carry specific loads point-to-point to inland regions in competition with rail, but there would a/so be a reduction of 1-5, 99 and 101 long haul transportation on traditionally diesel-fueled trucks over promotion of local and regional distribution via cleaner-fueled trucks such as LNG. Electric hybrid or future hydrogen power plants are currently deemed unfeasible for long haul, but possible for 120-mile radius operations. Main line and short rail opportunities would be enhanced under this plan as well, with an opportunity for clean or alternative fuel, or hybrid or electric operated systems.

The governments' role at all levels is to encourage innovation in these areas, including direct research, development, and practical support for the testing of concepts. LLNL; transportation centers such as those located at SJSU, CSU Long Beach, and UC Davis; the National Transportation Research Board; and maritime subject matter experts from the California Maritime Academy would be a well-qualified team to explore options and potentials if properly funded and supported by federal, state and local governments.

While all of these alternatives must be decades-long strategies to attain and realize the maximum public benefit through a deliberate planning process, elements of these programs are viable and ready to be implemented now, and should be funded to move the process forward.

Private Sector Recommendation #1"Advancing the Jones Act, While Considering Revisions That Can Strengthen Its Original Intent of Promoting the Us Maritime Industry and Protecting National Security Interests"

One of the major impacts to the U.S. shipping industry has been direct or indirect support of foreign flag carriers by flag state nations such as China, essentially resulting in the de facto ownership and control of otherwise commercial maritime assets. This situation is inherently unfair to commercial enterprises who are forced to compete with them. The Jones Act was envisioned to assure viability of coastwise trade in the waters of the United States and to assure a viable deep water capacity for national defense and national security military lift. With an increasing disappearance of the U.S.-flag oceangoing merchant fleet (currently less than 90 vessels), the ability of the United States to support rapid deployment of war material when and where needed is being significantly degraded and may well result in a future dependence on foreign flag charters for this critical logistical effort- a seriously flawed process. Historically, no nation in the world has maintained its global influence and supremacy without a strong and flexible Naval and Merchant Marine fleet. This status can only be maintained through a robust shipbuilding and repair infrastructure, exceptional workforce skills training and domestic employment, the highest level of professional mariner education (such as that provided by the six state chartered maritime academies), and a full commitment to exploration, development, and deployment of the next generation of efficient ships. All of these elements could be incorporated into a reinvigorated Jones Act for the maximum benefit of the nation and the maritime states. Attainment of these goals will provide the nation renewed control over our maritime borders, global goods movement, and implementation of effective foreign policy when and where maritime power projection and humanitarian assistance is needed.

Environment Sustainability Recommendation #3 "Congress & State Policy Makers to Consider the Following Principles When Developing New Environmental Policies Impacting the Maritime Industry"

California historically leads the nation in developing and implementing marine environmental protective standards - by some measures the toughest in the world. If its goals, which apply to eleven commercial ports and several private port operations within the state, are to be both socially desirable and technologically attainable, Congress and the appropriate executive agencies of government on the national level should look to streamline and standardize national standards applicable to all states to assure a level field for competitiveness and attainment of a healthy environment for America as a whole. Industry cannot be expected to independently fund 100 percent of federally-mandated changes without significant impacts to business viability and the national economy. Therefore, state and federal incentive programs should be considered to ease the burden and positively embrace required changes. If economic incentives are required to attain common standards, Congress should consider block grants directly to ports and their maritime stakeholders for the attainment and maintenance of common national environmental standards.

Education and Workforce Recommendation #2 "Easing Licensing Requirements and Programs for Veterans Entering the Maritime Workforce"

In consideration of the prevailing general circumstances, it is counterintuitive to expect the U.S. Navy and U.S. Coast Guard to invest millions of dollars training sailors, and then expect these organizations to fully embrace an easy exit program for the best of these men and women who are attracted to better working conditions, pay, and other incentives in the commercial maritime sector. Both organizations can and should consider obligation requirements, particularly for critical specialized skills, that encourage military personnel to remain in the service while also incentivizing them to stay as a fundamental way to attain an acceptable return on investment. Nevertheless, many sailors will choose to leave the maritime military services, and when they do, it is a national responsibility to assist these men and women to transition as efficiently and expeditiously as possible to commercial, civilian jobs where these perishable skills can be retained for the benefit of the nation. The nation's six state maritime academies can be enlisted to validate and certify basic skills using existing USCG-certified training courses, but the services themselves must be willing to certify practical sea time and PQS attainment. It is operationally and fiscally unwise and morally indefensible to install unnecessary barriers to the transition of qualified sailors from the military to the commercial sector where these skills can continue to be utilized for public benefit.

Education and Workforce Recommendation #3 "Work Together to Develop a New Era of Life-Saving Equipment"

The USCG establishes the rules and regulations for required life safety systems and equipment in the United States utilizing minimum baseline recommendations established by the International Maritime Organization (IMO). For international routes, the IMO standards are the minimum requirements. The USCG can- and often does - require more restrictive requirements for U.S.-flag vessels, but this process is often negotiated through alternative mitigation programs, policies, and procedures. It is clear that larger, state-of-the-art "blue water" ships require an equivalency in state-of-the-art life safety equipment, sized and sufficient to serve all crew members in reliable, enclosed survival craft deployed in adequate quantity and extreme weather conditions. In addition, review of life safety equipment for coastwise and inland vessel operators is overdue for review and upgrade. For example, passenger ferry operators are provided multiple alternative options to support the reduction of numbers and types of survival craft, as well as the required number of crew members in ratio to the numbers of passengers carried, to the extent that woefully inadequate practical safeguards often exist to counter the effects of a major marine casualty, should one occur. The continued authorization for use of Inflatable Buoyant Apparatus (IBA's) on coastal and inland ferries, in numbers inadequate to safely manage the high peak numbers of passengers and crews aboard, virtually assures significant numbers of passengers will end up in the water if a major incident occurs. Significant changes in these practices are realistically possible and absolutely necessary to protect the seafarer and the general public- before a major incident again makes clear the inadequacy of current practices.

A U.S. Merchant Marine recruitment poster from WWII. During WWII, the U.S. Merchant Marine had the tremendous responsibility of delivering troops, supplies, and equipment to Allied front lines.

The S.S. Red Oak Victory, *built in Richmond, California, saw duty during World War II, the Korean War, and the Vietnam War. Destined for scrap metal in Suisun Bay, she was saved by volunteers and an act of Congress and today is part of the Rosie the Riveter/World War II Home Front National Historical Park in Richmond. Of the 534 Victory-class cargo ships completed, she is one of only three that remains working, and she is the last surviving ship built at the Kaiser Richmond Shipyards during World War II.*

Appropriate regulations and oversight are important to preserving the natural beauty surrounding our shipping channels. A thriving maritime sector is compatible with protecting wetlands like the Suisun Marsh.

STATEMENT OF GENERAL DARREN W. MCDEW, UNITED STATES AIR FORCE COMMANDER, UNITED STATES TRANSPORTATION COMMAND, BEFORE THE HOUSE ARMED SERVICES COMMITTEE, READINESS SUBCOMMITTEE AND THE SEAPOWER AND PROJECTION FORCES SUBCOMMITTEE, ON THE STATE OF THE COMMAND, 30 MARCH 2017

Introduction

The United States Transportation Command (USTRANSCOM) delivers National objectives on behalf of the United States, and has proudly done so for nearly three decades. As we near our 30th anniversary, we stand ready to deliver an immediate force through our airlift and air refueling assets or a decisive force with our strategic sealift assets when and where needed. Our delivery of these forces assures an unparalleled global expeditionary capability and gives our Nation options when needing to respond to a variety of crises. Ultimately, this unmatched capability extends a helping hand or projects combat power anywhere, at any time and provides a key strategic advantage for our Nation. We must continue to invest in and preserve our edge. Our ability to sustain strategic power projection is challenged on several fronts by potential adversaries growing ever more capable. However, we continue to look forward and innovate as we face challenges, uncertainties, risks, and complex demands placed upon the unique capabilities we provide daily to our Nation.

Mission

USTRANSCOM delivers full-spectrum global mobility solutions supporting our Nation's requirements in peace and war. In the simplest terms, we provide viable national security options to the National Command Authorities. Those options range from immediate humanitarian or combat deliveries by our airlift and air refueling fleets, to the global delivery of the Nation's decisive combat power via our strategic sealift fleet. While ensuring the readiness and availability of these options on a daily basis, the command also leads the Joint logistics enterprise which is the foundation on which every other Department of Defense capability rides.

Although transportation remains USTRANSCOM's core competency, our span of influence extends from the source of supply, through each segment of the DoD supply chain to any designated point of need. Through logistics enterprise forums, the Services, Combatant Commands, DoD interagency partners and commercial providers, we continue to collaborate with logistics leaders in order to better identify, prioritize, and close capability gaps within the enterprise. Since 2009, these efforts have resulted in efficiencies and cost avoidances for the DoD distribution supply chain of nearly $1.6 billion, which translates to more buying power for the Services.

In addition to our primary mission of providing full-spectrum global mobility solutions, our subordinate command, the Joint Enabling Capabilities Command (JECC), provides decisive, rapidly deployable joint command and control capabilities. By doing so, they assist in the initial establishment, organization, and operation of joint force headquarters.

The JECC is a unique total force joint organization that delivers highly effective, cost efficient, joint planning, public affairs, and communications capabilities to all combatant commanders. This Subordinate Command is alert-postured to respond across the full range of military operations. They routinely deliver high-impact mission-specific teams of experts who produce executable solutions for emergent global crises. In fact, they provided more than 40,000 man-days of support that touched every combatant command in 2016; notably providing key Joint Task Force staff and planning expertise within 72 hours to assist Southern Command in responding to Hurricane Matthew. In addition, the JECC's robust support to the Chairman of the Joint Chiefs of Staff Combatant Command Exercise and Training program improved our National proficiency, knowledge, preparation, and response to emergent events.

The JECC continues to have significant forces deployed in support of missions around the globe, including direct support to ongoing counterterrorism operations in Iraq and Afghanistan. Their ability to deliver highly effective joint planning, public affairs, and communications capabilities have enabled better integration of DoD, U.S. government, and partner responses to strategic challenges in every part of the world.

Operating Environment

Today's diverse global security environment is dramatically different and more complex than the one we operated in for the last 30 years. China, Russia, Iran, North Korea, and violent extremist organizations continue to challenge us in new ways, ultimately demanding new and innovative solutions. As we look to the future, we expect trans-regional, multi-domain, and multi-functional conflicts will define our future operating environment. A global view is essential in such conflicts. Fortunately, USTRANSCOM has operated globally every day since its inception.

We also expect that future conflicts will cross regional boundaries and potential adversaries and peer competitors will field numerically superior forces with near-technological parity. Those adversaries are aware the United States has become accustomed to geographically- isolated conflicts and enjoyed technological superiority over its adversaries, so we expect contested global sea lanes and air routes to a degree we have not faced since World War II. Potential adversaries seek asymmetric means to cripple our force projection and sustainment capabilities by targeting critical military and civilian assets, both within the U.S. and abroad. Additionally, our enemies continue to use our dependence on the cyber domain against us. With those challenges in mind, every Soldier, Sailor, Marine, Airman, Coast Guardsman and Department of Defense (DoD) Civilian of USTRANSCOM and its Component and Subordinate Commands recognizes it is our duty to ensure the Command remains postured to operate in such an environment and effectively answer the Nation's call, should it come.

Our Approach

To effectively operate in that future environment, we are focused on four priorities: ensuring readiness today while advocating for future capabilities, advancing our capabilities in the cyber domain, evolving the command for tomorrow, and championing an innovative, diverse, and agile workforce.

In prioritizing today's readiness while keeping an eye toward future capabilities, we are advocating for the right mix of personnel, platforms, systems and training to ensure we can provide the global transportation and logistics capabilities our Nation requires. By making the right investments today in enhancements for our air, sea and surface fleets, we will ensure USTRANSCOM delivers the Nation's objectives tomorrow. Through this pursuit, we posture to meet the full range of Unified Command Plan roles and missions in current and emerging trans- regional transportation, logistics, and patient movement requirements.

Improving our cyber defense allows USTRANSCOM to operate freely and effectively. We continue to broaden our scope to actively evaluate and mitigate our command and control, weapon system, and infrastructure vulnerabilities, while identifying and advocating for the critical capabilities, policies, and procedures that ensure mission accomplishment.

As a global Combatant Command charged with delivering national objectives in tomorrow's dynamic security environment, we must challenge our assumptions, accurately forecast trends that shape that future environment, and develop the technologies and ideas that maintain our Nation's competitive advantage. In our pursuit to continuously evolve for tomorrow, we established a relationship with the Defense Innovation Unit Experimental to expedite the implementation of logistics- and cyber-based technologies such as commercial cloud-based technologies which can provide Infrastructure-as-a-Service, Platform-as-a-Service, and Software-as-a-Service offerings to host USTRANSCOM's unique applications. These and other initiatives will increase our responsiveness, agility, efficiency, and operational processes, while enhancing transparency with our customers and positioning us ahead of emerging challenges and threats.

Finally, our approach recognizes USTRANSCOM requires a talent rich, diverse, creative, adaptive, and innovative workforce to survive today and thrive in tomorrow's dynamic environment. We will create this workforce by recruiting, developing, and retaining the best talent America has to offer. We recognize that doing so requires us to remove cultural, procedural, and policy barriers along the way such as significant civilian hiring reform. We appreciate the attention Congress has placed on this issue with the recent passage of legislation in the 2017 NDAA creating a streamlined civilian on-campus recruiting authority, fast tracking the ability to hire talented personnel for critical positions in an ever-more competitive marketplace. Initiatives improving the speed at which talent can be hired, and opening aperture to additional fast-tracked hiring authorities and policy flexibility, ensures better access to streams of talent benefitting USTRANSCOM.

State of Our Readiness

Without reservation, USTRANSCOM stands ready to deliver on behalf of the Nation today. However, as our approach to the future operating environment indicates, there are challenges that demand our attention to ensure our readiness is never called into question. These challenges fall into the following broad categories: airlift and air refueling, sealift, surface, budget, and workforce issues.

Airlift/Aerial Refueling

Air Mobility Command (AMC), a Component Command of USTRANSCOM, provides an incredible capability to our Nation and the world. As one Total Force team with commercial partners, AMC provides airlift, aerial refueling, air mobility support, and aeromedical evacuation around the globe, supporting eight combatant commands while operating in 23 countries.

On average, tankers are conducting aerial refueling operations every five minutes over the skies of Iraq and Afghanistan. Additionally, AMC refuels fighter squadrons across the Pacific Ocean to ensure a constant presence throughout the Pacific and refuels nearly all of our North Atlantic Treaty Organization allies during operations and exercises around the globe. Yet the scarcity of forces and their current distribution, coupled with the high operations tempo placed upon them, comes at a cost to the health of the KC-10 and KC-135 fleets.

Currently, the KC-46A program is on track to deliver 179 aircraft by 2028, which will enhance operational agility. The delivery of these aircraft over the next few years remains a critical investment to ensure we can continuously project power around the world, whether in support of humanitarian relief missions or combat operations.

The delivery of the KC-46A alone, however, will not address present concerns with the allocation and distribution of the global tanker fleet. For example, since 2011, the authority of the USTRANSCOM commander to manage tankers globally has been constrained by congressional language prohibiting changes to command and control of scarce KC-135 forces. As global tanker requirements continue to expand, we are seeing more cases where the requirements of a given combatant command are in competition with others. At the same time, we recognize a trans-regional, multi-domain, and multi-functional operating environment combined with proliferating anti-access and aerial denial threats will only place greater strain on the air refueling force. Although we continue to work closely with AMC and the Joint Staff to mitigate the effects on the global tanker fleet, the restriction in place since 2011 continues to limit the USTRANSCOM commander's ability to exercise operational control of high demand Pacific and European tanker forces necessary to meet global and national defense requirements.

Also key to air mobility are our airlift assets and the additional capabilities commercial industry brings to the fight. For our part, the C-17 and C-5 continue to provide strategic airlift the world over while our C-130s meet tactical airlift needs in every region of the globe. We are seeing stress on the strategic airlift fleets and have some concerns about hard choices that have been made to close active duty C-17 squadrons with an eye toward buying that capability back in the reserve component. While both components are capable, maintaining the right balance is critical so we do not create a situation where mobilization is needed for every new mission that might arise.

On the commercial airlift side, our Civil Reserve Air Fleet (CRAF) partners commit aircraft to augment DoD airlift during contingencies and/or emergencies in exchange for government airlift business while also providing commercial airlift services to DoD during peacetime. USTRANSCOM closely coordinates with the Department of Transportation (DoT) in administering the CRAF program. This steadfast relationship has historically provided lift for roughly 40 percent of all DoD air cargo and 90 percent of all passenger movements in direct support of our warfighters. Our 24 CRAF carriers remain ready to support DoD readiness requirements with cargo and passenger support worldwide and we will continue to rely on viable and healthy CRAF program in the future. To ensure the relationship with our CRAF partners remains robust, we've begun contracting with them based on early demand signals, allowing us to move workload to them which would have otherwise been handled by our organic aircraft. This has the dual benefit of providing additional workload to our CRAF partners while also reducing flying hours in our organic fleet.

Sealift

Historically, nearly 90 percent of wartime transportation requirements are delivered through strategic organic and U.S. flagged commercial sealift. In fact, our strategic sealift fleet provides the ability to deliver a decisive force over great distances. Our U.S. Navy component, the Military Sealift Command (MSC), provides sealift capabilities through ship chartering, prepositioning, and sustainment operations while also executing operational command over the Maritime Administration's (MARAD) Ready Reserve Force ships during contingencies. Without a healthy and viable U.S. Commercial Sealift Fleet, MSC Surge Fleet, and MARAD's Ready Reserve Force, our Nation's military may not be able to deploy as quickly and efficiently as it can today.

The National Security Directive on Sealift and the Sealift Emergency Response Programs provide assured access to U.S. flagged commercial sealift assets, Merchant Mariners, and the global intermodal capability required to augment government owned (organic) sealift capabilities during contingencies. This assured access is provided via the Voluntary Intermodal Sealift Agreement (VISA) which ultimately ensures the U.S. maintains its capability to meet sealift requirements in peace, crisis, or war. The VISA program provides a responsive transition from peace to contingency operations through pre-coordinated agreements for U.S. flagged commercial sealift capacity and systems to support DoD's contingency sealift needs. It allows USTRANSCOM to meet mobilization requirements in a rapid fashion.

At the core of the VISA vessels are our Maritime Security Program (MSP) partners, who are essential to our wartime U.S. commercial sealift capability, and all are participants Sealift Emergency Response Programs. Over time, MSP has provided access to required commercial U.S. flag shipping assets, while also supporting the pool of Merchant Mariners needed to operate MSC's Surge and Ready Reserve Fleet. In this way, the MSP significantly contributes to the supply of Merchant Mariners available to serve on U.S. vessels in time of war while mitigating future risk to our national commercial capacity.

Along with MSP, The Merchant Marine Act of 1920, also known as the Jones Act, provides an additional pool of trained Merchant Mariners and sealift capacity. It does this, and contributes to national defense, by subsidizing a robust, domestic, maritime industry including U.S. industrial shipyard infrastructure for building, repairing, and overhauling U.S. vessels.

Ensuring a healthy U.S. fleet has proven difficult in the larger global context where international shipping has slowed while the industry as a whole has ended up with an excess of ships. Excess supply has caused prices to fall, which has put considerable financial pressure on U.S. flagged vessels. Unfortunately, the U.S. flagged international commercial fleet and Mariner pool has shrunk over time; while we have contingency plans, further reductions may cause us to investigate other options such as using more foreign flagged international commercial vessels manned by foreign crews during crisis or war. American shipping companies continue to re-flag vessels to foreign nations, diminishing the size of our commercial fleet, although that fleet stabilized in recent years at around 80 today. While the U.S. flagged commercial fleet remains the most effective means for us to obtain the necessary sealift capability to meet national defense needs, we are considering a range of options to ensure that we retain the ability to deploy a decisive combat force at the time and place of our choosing. Those options may include new approaches to preserving essential capabilities in the Ready Reserve Force, which among other options, may include non-US built vessels.

Sealift Fleet Recapitalization

Since the 1990s, DoD mobility studies have indicated a requirement for nearly 20 million square feet of Roll-on/Roll-off (RO/RO) capacity to promptly transport materiel wherever needed in defense of the Nation's interests in major conflicts abroad. This includes over 15 million square feet of organic RO/RO capacity on 65 total ships and nearly 4.5 million square feet of U.S. flagged commercial RO/RO capacity gained through VISA. However, we are projecting a loss of 4.5 million square feet of organic surge RO/RO capacity by 2033 as these vessels reach the end of their service life, with an accelerated loss expected between 2026 and 2031. The organic vessels tasked to meet this requirement are becoming obsolete or unsustainable. Our organic surge vessels, for example, have an average age of 39 years and will begin to reach their 50-year service life in the 2020s. This will result in a critical sealift capacity shortfall, which limits our ability to support the national security requirements. In addition to the RO/RO capacity loss, 10 of 12 special-capability ships will age out of the fleet between 2020 and 2024. These ships provide expeditionary capabilities such as over-the-shore fuel distribution and crane lift to austere or damaged ports, a critical necessity for the deployment of ground forces and for operations in a contested environment. The aging and loss of sealift capacity places a particular urgency on the need to explore options for maintaining critical capabilities, without which the Nation's strategic sealift capability to support future operations will be at risk. We are working closely with the U.S. Navy to maintain the full spectrum of strategic sealift capabilities required to move U.S. forces in current and future operational environments.

In order to keep the recapitalization strategy on track and achieve success in the near term, the used vessel acquisition component must start as early as fiscally possible. This component seeks to purchase vessels leaving MSP or other commercial vessels regardless of country of origin. The acquired vessels would replace the aging organic vessels for a fraction of the cost of new construction and could remain in service for several decades.

Congressional support will be needed to gain the necessary authorities and funding for this effort.

Additionally, DoD's current organic surge fleet is composed of several steam-propelled ships. The manning of these ships with seasoned steam certified engineers is a growing concern as commercial industry is expected to retire all steam ships by the early 2020s, while we need to operate them until 2035 unless recapitalization efforts allow us to replace them sooner. As commercial industry retires their steam ships, our access to a civilian pool of steam-certified engineers and mariners may fall severely (and eventually be) eliminated.

Surface

Since the first stage in delivering a decisive force happens on the ground, our Nation's infrastructure of roads, rails, and ports plays a fundamental and crucial role in the deployment and sustainment of the Joint force. USTRANSCOM closely partners with the Department of Transportation (DoT) and other Federal and State entities to ensure infrastructure within the continental U.S. is ready to support DoD deployment and distribution needs.

USTRANSCOM, through its Army component the Military Surface Deployment and Distribution Command (SDDC) represents the interests and requirements of the DoD to access and safely utilize both private and public transportation infrastructure and services. Currently, the public sector road network remains capable of meeting DoD ground transportation needs while providing adequate access to commercial trucking capacity to meet current and anticipated surface transportation needs.

Just as the availability and safety of drivers and roadways are critical to national defense, our national rail system is of equal importance. Through our Railroads for National Defense Program, and close collaboration with civil sector rail officials and DoT's Federal Railroad Administration, we assess the ability of the U.S. rail system to support military needs. Currently the rail network required to deploy our force is in place and viable. In the next fifteen years, however, we face age-mandated retirements of some of our uniquely capable DoD railcars. We are developing a plan in close collaboration with the Department of the Army to retain this critical transportation capability.

To successfully execute our deployment mission, USTRANSCOM also relies on a collection of both DoD and commercially-owned U.S. seaports, designated as Strategic Seaports.

The primary mission of the Strategic Seaport Program is to ensure DoD has access to sufficient seaport infrastructure to meet contingency deployment needs. None are more important than the main West and East Coast ports of Military Ocean Terminal Concord (MOTCO) and Military Ocean Terminal Sunny Point (MOTSU).

MOTCO on the West Coast is indispensable to USTRANSCOM's support of U.S. Pacific Command's operations and DoD's military capability in the Pacific Theater. Due to the nature and size of this mission, no suitable alternatives exist on this coast and MOTCO's infrastructure assets require critical upgrades and maintenance to remain relevant in the current military environment. Current efforts are centered on preserving the operability of MOTCO's primary pier until it can be replaced. We are also examining additional options for MOTCO to become a modern ammunition port, fully capable of safe and efficient operations to enable uninterrupted delivery of ammunition to the Pacific theater.

At MOTSU, significant infrastructure improvements within the last few years have enhanced our ability to support the EUCOM, AFRICOM, and CENTCOM Combatant Commanders' operations and allow the terminal to meet throughput requirements. We continue to work with the Department of the Army to preserve both seaports by finding and applying resources to reduce risk and prevent capability gaps.

Although our nation's roads, rails, and ports play a fundamental role in the deployment and sustainment of our Armed Forces, nothing is more important than the people and their families who support and execute the mission, at home and abroad. The Defense Personal Property Program provides our Service members, DoD Civilians, and their families with an effective and efficient system for the relocation, storage, and management of their household goods and privately-owned vehicle (POV) shipments. As a testament to our partnership with the Deputy Assistant Secretary of Defense for Transportation Policy, the Services, and commercial transportation service providers, the Defense Personal Property Program enables nearly 875 service providers to support the movement of approximately 70,000 POVs and around 430,000 household goods shipments globally each year.

An initial review and analysis of the Defense Personal Property Program identified that the end-to-end household goods value chain lacked proper alignment to consistently produce high quality relocation services due to lack of a single program manager and the need for more commercial capacity. Consequently, stakeholders from across the Department have collaborated on initiatives to recommend to key leaders from the military Services. Additionally, the Defense Digital Service recently assessed our primary software and customer interface within the Defense Personal Property System and found that while we are working toward reasonable solutions to the problems our customers have with the system, we are plagued by common problems that can be overcome by focused effort and assistance from the Defense Digital Service. Although still a work in progress, these initiatives are intended to improve the move experience for our Service members while simultaneously posturing the program for institutionalized and sustained continuous improvements. Addressing the challenges of program accountability, customer service, entitlements, standardization, and automation provides an opportunity to incorporate updated technologies and processes that will improve the quality of life and security of our Service members, DoD Civilians, and their families.

Budget Uncertainties

USTRANSCOM's mobility readiness depends highly on our financial health, which in turn relies on the financial posture of Services that provide the capabilities we use to execute our critical missions. Notably, the Budget Control Act and recent Continuing Resolutions have forced the Services to prioritize immediate operational needs over prudent long-term planning and investment, decisions that ricochet inefficiencies through the logistics and transportation enterprises. More generally, sequestration's impact on readiness, mission operations, and modernization funding will result in the Services being unable to adequately man, train, and equip mobility forces thus presenting a significant readiness challenge. Irregular budgets jeopardize critical mobility acquisition programs like the schedule for the KC-46A program which addresses significant challenges with our aging aerial refueling fleet.

Additionally, our overall readiness and that of other combatant commands is influenced by the joint training and exercises conducted solely through resources provided by the Combatant Commander's Exercise Engagement and Training Transformation program. With an increased emphasis on trans-regional, multi-domain, and multi-functional operations, adequate support to combatant command joint training and exercise programs is as critical as ever.

As we head into FY18, we are projecting adequate financial levels to ensure our readiness, but remain vigilant in light of budget uncertainty. While maintaining our overall readiness, we continue to focus on our commitment to becoming audit ready by complying with the initiatives of the Financial Improvement and Audit Readiness (FIAR) program. At the same time, as we move to multi-modal operations and new Plan, Order, Ship, Track and Pay processes, we are integrating FIAR throughout those processes to ensure audit compliance. Furthermore, we continue to work closely with our DoD counterparts to ensure Transportation Financial Audibility throughout the Department. In an effort to ensure our business processes remain relevant today and into the future, we are striving to be cost-competitive and more transparent with our customers while simultaneously seeking ways to provide a lower, more predictable rate structure. These actions enhance our coordinating role across the deployment and distribution enterprise and ultimately enhance the support we provide the warfighter.

Availability of the Transportation Workforce

Each mode of our transportation network (air, surface, and sea) requires commercial and/or military operators such as truck drivers, aircraft pilots, and sealift mariners. These operators allow USTRANSCOM to transport forces and materiel to the point of need and to return our ill and injured to appropriate medical care. Worsening shortages of these operators limit our ability to successfully deliver required combat power across the globe.

In order to respond anywhere in the world in a matter of hours, appropriate manning levels of both Air Force and commercial pilots are essential. In fact, all DoD aerial refueling and nearly all strategic aeromedical evacuation capability relies on the availability of the U.S. Air Force aircrews from the active and reserve components. Additionally, USTRANSCOM's organic and commercial airlift capabilities deliver roughly 10 percent of all transportation requirements and continue to be a significant force multiplier for the Nation by delivering an immediate force overnight into an area of operations when needed. Pilot manning will remain vital for the near and long-term future of this critical USTRANSCOM capability.

Our Mobility Air Forces and commercial airline partners, however, are experiencing manning shortages. Reduced undergraduate pilot training quotas, changes in force structure, and declining retention (along with aggressive airline hiring for the foreseeable future) will require a concerted effort if we are to mitigate potential negative impacts across the active and reserve components of the U.S. Air Force. By comparison, U.S. commercial airlines, including our partners participating in the CRAF program, expect a pilot shortage of roughly 35,000 pilots through the year 2031. Contributing factors include retirements exacerbated by statutory age limits (i.e., max of 65 years old), an increase in new airline transport pilot certificate requirements, and the continuous growth of the global airline industry.

While monitoring the health of pilot manning, we are keenly focused on the strain the trucking industry continues to feel due to the commercial truck driver shortage (current shortage of 75,000 with a projection of 890,000 by 2027). The industry's ability to attract and retain qualified commercial truck drivers required to move freight for the DoD and the Nation is a growing concern. Persistent shortages are caused by several factors such as quality of life, younger generations not seeking out the truck driving profession, and the pursuit of more desirable job alternatives. Due to the shortage of operators, the trucking industry is currently operating at greater than 95% capacity leaving little to no surge capacity for DoD. In an effort to address and reverse the shortage of drivers, the DoT Federal Motor Carrier Safety Administration implemented initiatives such as granting test waivers for military members with previous commercial motor vehicle experience and expanding their experience validation timeline from 90 days to one year. These efforts are the first steps in addressing the shortage; however, they may not be sufficient to alleviate likely impacts in times of increased demand for DoD's surge requirements. To combat this issue, we are continuously engaged with industry and conducting ongoing analysis and reviewing plausible alternatives, such as increasing the use of multiple modes (rail and truck) and un-manned vehicles.

We will continue to monitor and manage the manning shortages across our three fundamental domains. While supporting DoT's efforts in reversing current trends, we remain determined to ensure a sufficient pool of transportation operators are available to provide our Nation transportation options.

Advancing the Cyber Domain

The greatest challenge USTRANSCOM faces every day is the threat of attack from the cyber domain. Although cybersecurity is a DoD-wide focus area, USTRANSCOM is distinctly vulnerable because the majority of the Command's transportation data resides within and travels through the unsecure commercial internet. Furthermore, unsecure networks and systems of our commercial transportation service providers, coupled with critical infrastructure vulnerabilities around the globe, almost wholly reside outside our control and pose significant risk to mission assurance. Due to these challenges, USTRANSCOM is prioritizing our key cyber concerns. The Command is collaborating with U.S. Cyber Command, DoD Agencies, Federal cyber organizations, industry, and academia to identify and mitigate gaps and shortfalls, as well as to seize opportunities to advance our cyber domain capabilities.

We have and will continue to encourage industry partners to join together in a transportation-related Information Sharing and Analysis Organization, focused on supporting DoD's transportation mission. Our current partnership with the National Defense Transportation Association (NDTA) Cybersecurity Committee and our semiannual Cybersecurity Roundtables that have drawn interest from across government, industry, and academia are cornerstones of our efforts to build a more responsive, aware, and collective approach to mission assurance.

For example, through our partnership with the NDTA Cybersecurity Committee, we seek to address: cybersecurity issues of mutual concern, rapid sharing of threat information, the application of best practices, and, research on existing and emerging cybersecurity technology and development activity. The committee provides a mechanism to address urgent concerns, such as cybersecurity contract language and the exchange of unclassified and sensitive information between USTRANSCOM and industry partners.

However, there is still much more to do in order to address our current and future cyber capabilities. People, processes, and technology are all key areas where we can enhance our cyber resiliency. Specifically, we must grow and retain a highly skilled cyber workforce; clarify cybersecurity roles and responsibilities across the critical infrastructure sector; implement acquisition policy to provision cloud services and other innovative cyber solutions at the "speed of need." We also need to ensure that DoD can rapidly share threat and vulnerability information with our commercial partners while continuing to improve cybersecurity compliance reporting that enables informed risk management decisions appropriate to the situation at hand. We will continue to work with U.S. Cyber Command, DoD Agencies' cyber organizations, Department of Homeland Security, and commercial transportation partners to mitigate cyber risks to global distribution operations.

Evolving for Tomorrow

Beyond the contested cyber domain, we recognize we will face new challenges across the other domains as well. As mentioned, we expect future conflicts are increasingly likely to occur in an environment contested across all domains, subsequently restricting our freedom of action. Adversaries and geopolitical competitors have the ability today to challenge our freedom of movement from deployment to employment using kinetic and non-kinetic means to disrupt, delay, or deny operations. This creates an environment that places our strategic assets at great risk. As our Joint force prepares to face this challenge, operational plans must reflect the anticipated attrition of both combat and mobility assets and associated personnel.

Future conflicts within the contested environment will also greatly challenge global patient movement operations. USTRANSCOM currently operates the best patient movement system in the world, safely and efficiently moving thousands of our nation's ill and injured Service members to the medical care they need every year. We are not content, however, to rest on our successes. Recognizing future adversaries might be able to limit our access to the air and cyber domains, we are aggressively exploring surface movement solutions to ensure we remain the best in patient movement no matter the threat or environment. We continue to partner with the Services to expand maritime patient movement capabilities and we are working to rebuild our ability to move patients by rail. Our Nation's joint casualty stream must be supported by joint patient movement capabilities that are interoperable, multi-modal, and capable of functioning in a cyber-compromised environment. Synchronized policies, training, and research and development are needed across the DoD to ensure we remain the best in safely moving our ill and injured whenever and wherever needed.

We also conducted a Future Deployment and Distribution Assessment focused on deployment and distribution in a contested environment. This assessment solidified our concerns about the challenges of conducting operations in contested environments and again highlighted that our global network (to include partners and allies) is at risk from threats in all domains – land, sea, air, space, and cyberspace.

Finally, we hosted our inaugural contested environment war game aimed at creating a common understanding of our operations in contested environments. We also sought to recognize the enterprise-wide challenges and develop prioritized mitigation efforts to enable future operations in those environments. The war game also addressed the necessary investments in planning and collaboration with the entire DoD logistics enterprise necessary to develop appropriate mitigation strategies for these threats. Finally, the war game highlighted the need for multiple operating options to ensure resiliency, agility, and responsiveness in future conflicts. An important insight from the war game is that operational plans and fleet sizing considerations must account for the loss of capital assets. These vital principles ensure realistic planning and aligned with risk, resulting in operational resiliency across all domains in future contested environments.

Based on these findings and published defense guidance, we will work with the Office of the Secretary of Defense, CAPE to initiate a new mobility requirements study once defense strategic guidance and the supporting elements, such as defense planning scenarios, mature.

Championing an Innovative, Diverse, & Agile Workforce

In light of all of the challenges and opportunities ahead of us, we recognize our people are our greatest resources and are the ones who will rise to those challenges and seize the opportunities. At the same time, it is clear the competition for talent is becoming more acute. Given that, we are working to recruit, develop, and retain the best talent America has to offer. At the heart of that effort we completed our first-ever headquarters Human Capital Strategic Plan setting short and long term goals for cultivating and managing our large professional civilian workforce. We also entered into an agreement with Defense Logistics Agency that created an individual civilian experiential development opportunity. Goal of effort is to address how USTRANSCOM and Defense Logistics Agency can work together to collectively develop the skills, knowledge, and effectiveness of our civilian workforce. We expect this initial agreement to serve as a baseline we can expand upon in the future. Beyond these early accomplishments, we are positioning ourselves to remain competitive in attracting future talent by identifying and establishing developmental positions within USTRANSCOM.

Our Commitment

For nearly three decades, our Nation has turned to USTRANSCOM's strategic power projection capability to respond rapidly to global threats and disasters. Today, USTRANSCOM continues to deliver 21st century, enterprise-wide, global expeditionary capabilities to the joint force. Anticipating and adapting to challenges will allow us to perform our missions in an ever- changing security environment. These missions continue to trend toward non-permissive, remote, austere, and widely-dispersed locations, but this team of transportation and logistics professionals always finds a way to deliver our national objectives. Our continuous focus on the resiliency and preservation of the Joint logistics enterprise while advocating for the right investments in our cyber-enabled air, sea, and surface fleets ensures we can deliver the Nation's objectives tomorrow.

Continued Congressional support, coupled with the hard work of the professional men and women of USTRANSCOM and our components, will ensure we are ready to deliver the Nation's Objectives. We will continue to address challenges and vulnerabilities and advocate for innovative solutions as we provide the joint force options for delivering an immediate force tonight and a decisive force when needed. "Together, we deliver!"

In: Coast Guard Programs
Editor: David J. Simmons
ISBN: 978-1-53615-426-9

Chapter 3

REVIEW OF THE FISCAL YEAR 2019 BUDGET REQUEST FOR THE COAST GUARD AND MARITIME TRANSPORTATION PROGRAMS*

Subcommittee on Coast Guard and Maritime Transportation

SUMMARY OF SUBJECT MATTER

March 9, 2018

TO: Members, Subcommittee on Coast Guard and Maritime Transportation

FROM: Staff, Subcommittee on Coast Guard and Maritime Transportation

RE: Hearing on "Review of Fiscal Year 2019 Budget Request for the Coast Guard and Maritime Transportation Programs"

PURPOSE

The Subcommittee on Coast Guard and Maritime Transportation will hold a hearing on Wednesday, March 14, 2018, at 11:00 a.m., in 2167 Rayburn House Office Building to examine the fiscal year (FY) 2019 budget requests for the Coast Guard and Maritime Transportation Programs. The Subcommittee will hear testimony from the U.S. Coast Guard (Coast Guard or Service), the Federal Maritime Commission (Commission or FMC), and the Maritime Administration (MARAD).

---

* This is an edited, reformatted and augmented version of Hearing before the Subcommittee on Coast Guard and Maritime Transportation of the Committee on Transportation and Infrastructure, House of Representatives, One Hundred Fifteenth Congress, Second Session, dated March 14, 2018.

BACKGROUND

Coast Guard

The Coast Guard was established on January 28, 1915, through the consolidation of the Revenue Cutter Service (established in 1790) and the Lifesaving Service (established in 1848). The Coast Guard later assumed the duties of three other agencies: the Lighthouse Service (established in 1789), the Steamboat Inspection Service (established in 1838), and the Bureau of Navigation (established in 1884).

Under Section 2 of Title 14, United States Code, the Coast Guard has primary responsibility to enforce or assist in the enforcement of all applicable federal laws on, under, and over the high seas and waters subject to the jurisdiction of the United States; to ensure safety of life and property at sea; to carry out domestic and international icebreaking activities; and, as one of the five armed forces of the United States, to maintain defense readiness to operate as a specialized service in the Navy upon the declaration of war or when the President directs.

The Coast Guard is directed by a Commandant, who is appointed by the President with the advice and consent of the Senate to a four-year term. On May 30, 2014, President Obama appointed Admiral Paul F. Zukunft as Commandant of the Coast Guard. His term expires in May 2018. On March 8, 2018, President Trump announced his intention to nominate Vice Admiral Karl Schultz as the 26th Commandant of the Coast Guard.

Fiscal Year 2019 Coast Guard Budget Request

The President requests $10.93 billion in FY 2019 for the activities of the Coast Guard. The FY 2019 request was in line with the spending caps for FY 2019 that were set in the Balanced Budget and Emergency Deficit Control Act. The Bipartisan Budget Act of2018 increased defense and non-defense discretionary spending caps in FY 2018 and FY 2019. Based on those increases, the President modified the FY 2019 budget request to reflect the increased cap levels through an Addendum which included $720 million for the Coast Guard for the construction of the first Heavy Polar Icebreaker. With this additional funding, the total request for the Coast Guard is $11.65 billion, which is an increase of $977 million above the FY 2018 request level of$10.673 billion, and an increase of$979 million over the FY 2017 enacted level of $10.671 billion.

Of the $11.65 billion requested in FY 2019, $9.7 billion is for Coast Guard discretionary accounts, $1.1 million (or 11 percent) more than the current FY 2017 enacted level. This amount does not include a transfer of approximately $160 million in funding to the Coast Guard from the Department of Defense (DoD) Overseas Contingency Operations account. The transfer of these funds would support the ongoing deployment of six 110-foot Coast Guard Patrol Boats conducting port and waterways security operations in the Persian Gulf.

The Coast Guard is currently operating under *The Continuing Appropriations Act, 2018* (P.L. 115-56, as amended by P.L. 115-123), that provides $4.244 billion (or 60 percent of the FY 2017 enacted level) in discretionary funding through March 23, 2018.

In FY 2019, the Coast Guard will transition to the DHS Common Appropriations Structure (CAS). Accordingly, activities funded through the previous Operating Expenses, Reserve Training, Environmental Compliance and Restoration, and Medicare-Eligible Retiree Health Care Fund Contribution are included as part of the new Operations and Support (O&S) account. In addition, acquisition personnel costs previously funded through the Acquisition, Construction, and Improvements account are included as part of the O&S account. The Acquisition, Construction, and Improvements account transitions into the Procurement, Construction, and Improvements account and the Research, Development, Test and Evaluation account becomes the new Research and Development account.

Program	**FY2017 Enacted**	**FY2019 President's Budget Request**	**Diff. Bet. FY2019 Budget Request & FY 2017 Enacted**	**% Diff. Bet. FY 2019 Budget Request & FY 2017 Enacted**
Operations Support[1]	$ 7,218,059	$ 7,792,498	*$ 574,439*	*8.0%*
Operating Expenses	$ 6,916,936	$ 7,462,054	*$ 54,118*	*7.9%*
Environmental Compliance & Restoration	$ 13,315	$ 13,429	$ *114*	*0.9%*
Reserve Training	$ 112,302	$ 117,655	*$ 5,353*	*4.8%*
Medicare-Eligible Retireee Health Care Fund	$ 175,506	$ 199,360	*$ 23,854*	*13.6%*
Procurement, Construction & Improvements	$ 1,370,007	$ 1,886,750	$ 516,743	37.7%
Research & development	$ 36,319	$ 19,109	$ (17,210)	*-47.4%*
Subtotal, Discretionary	$ 8,624,385	$ 9,698,357	$ *1,073,972*	*12.5%*
Retired Pay	$ 1,666,940	$ 1,734,844	$ *67,904*	4.1%
State Boating Safety Grants	$ 113,049	$ 114,682	$ *1,633*	*1.4%*
Maritime Oil Spill Program	$ 101,000	$ 101,000	$ -	*0.0%*
General Gift Funds	$ 2,829	$ 2,864	$ *35*	*1.2%*
Subtotal, Mandatory	$ 1,883,818	$ 1,953,390	$ *69,572*	*3.7%*
Total	$ 10,508,203	$ 11,651,747	$ 1,*143,544*	*10.9%*

[1] The FY 2019 President's Budget Requests combines 4 legacy appropriations into "Operations & Support."

Above is a comparison of the President's FY 2019 budget request to the FY 2017 enacted funding level.

Operations and Support (Previously Operating Expenses)

The President requests $7.79 billion for the O&S account in FY 2019, $876 million (or 11 percent) more than the FY 2017 enacted level. The O&S account supports the day-to-day activities of the Coast Guard including administrative expenses, support costs, travel, lease payments, and the operation and maintenance of infrastructure and assets. The O&S account also funds personnel compensation and benefits for the Service's approximately 41,000 active duty military members, 7,500 reservists, and 8,500 civilian employees.

The O&S budget includes increases in funding to cover follow-on costs for the operation and maintenance of newly acquired assets and technology and increases in other administrative expenses. The request includes a $275.2 million increase to cover the cost of the FY 2019 military pay raise (2.6 percent), as well as expanded military benefits enabling Coast Guard servicemembers to maintain parity with benefits received by DoD servicemembers, operational adjustments, and operating and maintenance funds for new assets.

Environmental Compliance and Restoration (EC&R)

The President requests $13.4 million for the EC&R in FY 2019, $114,000 (or 0.86 percent) more than the FY 2017 enacted level. The EC&R funding provides for the clean-up and restoration of contaminated Coast Guard facilities, as well as for the remediation of Coast Guard assets to ensure they are safe to operate or can be decommissioned in compliance with environmental laws.

The Coast Guard plans to use the $13.4 million requested for EC&R to pay for the continued long-term monitoring at 29 sites and begins or continues investigation/ remediation site work at 19 sites.

Reserve Training

The President requests $117.65 million for the Reserve Training in FY 2019, $5.4 million (or 4.77 percent) more than the FY 2017 enacted level. Reserve Training funding supports the costs of training members of the Coast Guard Reserve and the administration of the Reserve Program.

The FY 2019 request retains the FY 2017 enacted level of 416 positions. The FY 2019 request of 409 FTEs is a reduction of seven FTE from the FY 2017 enacted level of 416 FTEs. This request supports the readiness of the 7,000-member Coast Guard Reserve workforce.

Reservists maintain readiness through regular training and exercises. Reservists can be mobilized by the Secretary of Homeland Security to support the response to a national emergency or disaster, and by the Secretary of Defense to support national security operations worldwide. In 2017, Coast Guard Reservists were mobilized to support U.S. states and territories affected by Hurricanes Harvey, Irma, and Maria.

O&S increases are offset by $120.6 million in cuts derived through decommissioning certain assets, the elimination of a program and the associated military and civilian positions, and the termination of one-time costs. The proposed reductions in the O&S account include:

- *Reduction of Pharmacy Co-Pays:* In FY 2019, the Coast Guard anticipates a $7.3 million savings based on the FY 2018 National Defense Authorization Act (NDAA) proposal to modify retail and mail order pharmaceutical co-pays.
- *Termination of One-Time Costs:* The FY 2019 budget request proposes a $37.1 million savings associated with the termination of one-time costs for program start-up and exit transactions in FY 2018, including the termination of CG Aircraft FAA Compliance and Obsolete Equipment Replacement ($16.8 million) and National Security Cutter Follow-On ($7.7 million).
- *Elimination of Crew Rotation Concept Pilot Program:* The FY 2019 budget eliminates $31.7 million funding and 231 positions / 231 FTEs for the Crew Rotation Concept operations on National Security Cutter hulls 1-3 homeported in Alameda, CA.
- *Asset Decommissionings:* The FY 2019 budget proposes to decommission four HC-130H aircrafts, which are being replaced by the new HC-130J aircrafts. The Coast Guard estimates these decommissionings will save $16.2 million in FY 2019.

Procurement, Construction, and Improvements (Previously Acquisitions, Construction, and Improvements)

The President requests $1.89 billion for the Procurement, Construction, and Improvements (PC&I) account, a $516.7 million (or 37.7 percent) increase over the FY 2017 enacted level. The PC&I account funds the acquisition, procurement, construction, rebuilding, and physical improvements of Coast Guard owned and operated vessels, aircraft, facilities, aids-to-navigation, communications and information technology systems, and related equipment.

The FY 2019 budget request includes $1.76 billion for the acquisition of aircraft, vessels, and the continued build-out of Command, Control, Communications, Computer, Intelligence, Surveillance, and Reconnaissance (C4ISR) systems. This represents an increase of$597.1 million (or 51.7 percent) from the FY 2017 enacted level. The budget request includes:

- $30 million for the construction of a Heavy Polar Icebreaker. The FY 2019 Budget Addendum included an additional $720 million, for a total of $750 million;
- $65 million to conduct Post Delivery Activities on National Security Cutters (NSC) 7 through 9;
- $240 million for the production of four Fast Response Cutters (FRC);
- $400 million for the construction of the second Offshore Patrol Cutter (OPC) and to facilitate evaluation of the Long Lead Time Materials for OPC 3. The OPCs will replace the Service's aging 210-foot and 270foot Medium Endurance Cutters (MEC);
- $80 million to fund the requirement to establish logistics for 14 newly acquired HC-27J aircraft. The request funds HC-27J Asset Project Office activities, logistics, training, and engineering studies to assess and resolve aircraft obsolescence issues;
- $20 million for the continued modernization and sustainment of the HH-65 Dolphin helicopter fleet;
- $23.3 million for C4ISR design, development, and integration; and
- No funding for the Alteration of Bridges program in FY 2019. The program did not receive funding in FY 2017 or FY 2016. Established by the Truman-Hobbs Act of 1940 (33 U.S.C. 511 et. seq.), the Alteration of Bridges program authorizes the Coast Guard to share with a bridge's owner the cost of altering or removing privately or publicly owned railroad and highway bridges that are determined by the Service to obstruct marine navigation.

The budget requests $135 million to construct or renovate shore facilities and aids-to-navigation. This request is a $35.5 million (or 26.3 percent) increase over the FY 2017 enacted level. The Coast Guard currently has a backlog of 95 prioritized shore facility improvement projects with an estimated combined cost of over $1.5 billion.

Research and Development (Previously Research, Development, Test, and Evaluation)

The President requests $19.1 million in FY 2019 for the Coast Guard's Research and Development (R&D) account, $17.2 million (or 47.4 percent) less than the FY 2017 enacted level. The R&D account supports improved mission performance for the

Service's 11 statutory missions through applied research and development of new technology and methods.

The Coast Guard intends to use the $19.1 million in FY 2019 for programs to test and evaluate unmanned aircraft system (UAS) prototypes, continue development and testing of the next generation Arctic navigation safety information system, evaluate emerging maritime oil spill response technology, and evaluate existing cybersecurity tools for port critical infrastructure protection and resilience.

Coast Guard Fiscal Year 2017 Authorized Funding

On February 8, 2016, the President signed into law H.R. 4188, *the Coast Guard Authorization Act of2015* (P.L. 114-120). The *Coast Guard Authorization Act of2015* authorized funding for the discretionary accounts of the Coast Guard for FY 2017.

Below is a comparison of the President's FY 2019 budget request to the FY 2017 enacted authorization.

Program	**FY2019 President's Budget Request**	**FY2017 Enacted Authorization (PL 114-120)**
Operations and Support	$7,792,498,000	$6,981,036,000
Environmental Compliance & Restoration	$13,429,000	$16,701,000
Reserve Training	$117,655,000	$140,016,000
Procurement, Construction, & Improvements	$1,886,750,000	$1,945,000,000
Research & Development	$19,109,000	$19,890,000
Total	$9,698,357,000	$9,102,643,000

The *Coast Guard Authorization Act of 2017* is awaiting Senate action. The *Coast Guard Authorization Act of 2017* (passed by the House of Representatives as Division E of H.R. 2825) authorizes funding levels of$9.397 billion for FY 2018 and $9.587 billion for FY 2019. The FY 2018 and FY 2019 authorized levels were dictated by the funding caps in the *Balanced Budget and Emergency Deficit Control Act.*

Federal Maritime Commission

The Federal Maritime Commission (FMC or Commission) was established in 1961 as an independent agency that regulates oceanborne transportation in the foreign commerce of the United States. The FMC protects shippers and carriers from restrictive or unfair practices of foreign governments and foreign-flagged carriers. The FMC also enforces laws related to cruise vessel financial responsibility, to ensure cruise vessel operators have sufficient resources to pay judgments to passengers for personal injury or death or for nonperformance of a voyage.

The FMC is composed of five Commissioners appointed for five-year terms by the President with the advice and consent of the Senate. The Commission is led by a Chairman designated by the President. On January 23, 2017, the President designated Michael A. Khouri as Acting Chairman.

Fiscal Year 2019 FMC Budget Request

The President requests $27.49 million in FY 2019 for the activities of the FMC, $264,347 (or 0.97 percent) more than the FY 2017 enacted level. The Commission is currently operating under *The Continuing Appropriations Act, 2018 and Supplemental Appropriations for Disaster Relief Requirements Act, 2017* (P.L. 115-56, as amended by P.L. 115-123), that provides $13.02 million (or 52.2 percent of the FY 2017 enacted level) through March 23, 2018. Below is a comparison of the President's FY 2019 budget request to the FY 2017 appropriated funding.

The *Coast Guard Authorization Act of 2015* authorized the activities of the FMC at $24.7 million for FY 2017. H.R. 2593, as passed by the Committee, authorizes $28 million for FY 2018 and $28.5 million for FY 2019.

Account	**FY2017 Enacted**	**FY2019 President's Budget Request**	***Diff. Bet. FY 2019 Bequest & FY 2017 Enacted***	***% Diff. Bet. FY 2019 Request & FY 2017 Enacted***
Formal Proceedings	$8,273,972	$8,744,040	*$470,068*	*5.68%*
Equal Employment Opportunity	$146,544	$94,508	*($52,036)*	*-35.51%*
Inspector General	$367.331	$437.817	*$70,486*	*19./9"-1,*
Operational and Administrative	$18,437,806	$18,213,635	*($224,171)*	*-1.22%*
Total	$27,225,653	$27,490,000	*$264,347*	*0.97%*

Maritime Administration

The Maritime Administration (MARAD) was established in 1950. It administers financial programs to build, promote, and operate the U.S. flag fleet; manages the disposal of federal government-owned vessels; regulates the transfer of U.S. documented vessels to foreign registries; maintains a reserve fleet of federal government-owned vessels essential for national defense; operates the U.S. Merchant Marine Academy; and administers a grant-in-aid program for state operated maritime academies.

MARAD is led by an Administrator appointed by the President with the advice and consent of the Senate. On August 8, 2017, President Trump appointed Rear Admiral Mark H. Buzby, USN, Ret. to serve as Administrator.

Fiscal Year 2019 MARAD Budget Request

The President requests $396.4 million in FY 2019 for the activities of MARAD, $126.1 million (or 24 percent) less than the FY 2017 enacted level. The FY 2019 Budget Addendum included an additional $300 million for a one-time procurement and retrofitting of two used cargo ships to replace ageing State Maritime Academy training ships. With the addition of the $300 million, the total request for MARAD in FY 2019 is $696.4 million.

MARAD is currently operating under *The Continuing Appropriations Act. 2018 and Supplemental Appropriations for Disaster Relief Requirements Act, 2017,* that provides $247.4 million (or 52.7 percent of the FY 2017 enacted level) through March 23, 2018. Below is a comparison of the President's FY 2019 budget request to the FY 2017 appropriated funding.

Account	**FY2017 Enacted**	**FY2019 President's Budget Request**	***Diff. Bet. FY 2019 Request & FY 2017 Enacted***	***% Diff. Bet. FY 2019 Request & FY 2017 Enacted***
Operations and Training	$175,560,000	$452,428,000	*$276,868,000*	*157.71%*
Assistance to Small Shipyards	$1 0,000,000	$0	*-$10,000,000*	*0.00%*
Ship Disposal Program	$34,000,000	$30,000,000	*-$4,000,000*	*-11.76%*
Maritime Security Program	$300,000,000	$214,000,000	*-$86,000,000*	*-28.67%*
Account	FY2017 Enacted	FY2019 President's Budget Request	*Diff. Bet. FY 2019 Request & FY 2017 Enacted*	*% Diff. Bet. FY 2019 Request & FY 2017 Enacted*
Title XI - Administrative Expenses	$3,000,000	$0	*($3.000,000)*	*-100.00%*
Title XI - Loan Guarantees	$0	$0	*$0*	*0%*
Total	$522,560,000	$696,428,000	*$173,868,000*	*33.27%*

Operations and Training

The President's FY 2019 request of $152.4 million for Operations and Training (O&T) is $23.1 million less than the FY 2017 enacted level of$175.6 million. The FY 2019 Budget Addendum added $300 million in the O&T account as a one-time cost for the replacement of two of its aging Schoolships the TS *Empire State* and the TS *Kennedy.* O&T funds the salaries and expenses for each of MARAD's programs, the operation, maintenance, and capital improvements to the U.S. Merchant Marine Academy, and financial assistance to the six state maritime academies.

The FY 2019 budget request for O&T includes $74.6 million for the U.S. Merchant Marine Academy, including $70.6 million for Academy Operations, and $4 million for capital improvements, repairs, and maintenance; $24.4 million for the six state maritime academies, including $22 million for School Ship Maintenance and Repair; and $53.4 million for MARAD Operations and Programs. The budget does not request funding for the Marine Highways Grant Program.

Assistance to Small Shipyards

The FY 2019 budget does not request funds for the Assistance to Small Shipyards Grant Program. The program provides capital grants to small privately owned shipyards to expand and modernize shipbuilding capacity, efficiency, and competitiveness. The program received $10 million in FY 2017. The *Howard Coble Coast Guard and Maritime Transportation Act of2014* (P.L. 113-281) reauthorized the program through fiscal year 2017 at $10 million per year.

Ship Disposal

The FY 2019 budget requests $30 million for the Ship Disposal Program, $4 million (or 12 percent) less than the FY 2017 enacted level. The program provides for the proper disposal of obsolete government-owned merchant ships maintained by MARAD in the National Defense Reserve Fleet. This request includes $25 million to continue the decommissioning project for the dismantlement and decontamination of the defueled nuclear power plant on board the former Nuclear Ship SAVANNAH (NSS).

Maritime Security Program

The FY 2019 budget requests $214 million for the Maritime Security Program (MSP), $86 million (or 28.7 percent) less than the FY 2017 enacted level. Under this program, $214 million in direct payments are allocated among up to 60 U.S. flagged vessel operators engaged in foreign trade. MSP vessel operators are required to keep their vessels in active commercial service and provide intermodal sealift support to the DoD in times of war or national emergency.

Title XI Loan Guarantees

The budget does not request funds for new loan guarantees for the construction or reconstruction of U.S. flagged vessels in U.S. shipyards under the Title XI program. The program received $3.0 million in FY 2017. The President's FY 2019 budget request (similar to the FY 2018 request) proposes to the move the administration of the Title XI program from MARAD to the National Surface Transportation and Innovative Finance Bureau.

WITNESS LIST

Admiral Paul F. Zukunft
Commandant
United States Coast Guard

Master Chief Steven W. Cantrell

Master Chief Petty Officer of the Coast Guard
United States Coast Guard

Rear Admiral Mark H. Buzby, USN, Ret.
Administrator
Maritime Administration

The Honorable Michael A. Khouri
Acting Chairman
Federal Maritime Commission

REVIEW OF THE FISCAL YEAR 2019 BUDGET REQUEST FOR THE COAST GUARD AND MARITIME TRANSPORTATION PROGRAMS

Wednesday, March 14, 2018
House of Representatives,
Subcommittee on Coast Guard And Maritime
Transportation, Committee
On Transportation and Infrastructure,
Washington, DC.

The subcommittee met, pursuant to call, at 11:07 a.m., in room 2167 Rayburn House Office Building, Hon. Duncan Hunter (Chairman of the subcommittee) presiding.

Mr. HUNTER. The subcommittee will come to order.

Today we convene to review the fiscal year 2019 budget request for the Coast Guard and maritime transportation programs.

The United States Coast Guard is a multimission service and the only armed service with law enforcement authorities. I have ongoing concerns with the Coast Guard being an armed service within the Department of Homeland Security.

On its face, the Coast Guard should fit comfortably within the Department due to its role in defense and homeland security; however, under the Department, the Coast Guard doesn't fare well.

The Service is hampered by lackluster funding requests that don't meet the needs of the Service, and as we heard at last week's hearing, the Department restricts what performance measures the Coast Guard publicly reports.

Admiral Zukunft, during your 4 years as Commandant of the Coast Guard, we have seen the Service make progress reforming its acquisition process, acquisition programs for the National Security Cutter and the Fast Response Cutter. While they have had their issues in the past, they have been producing much-needed assets for the Coast Guard. The Offshore Patrol Cutter is coming online this year, and hopefully soon we will see the start of construction for a polar icebreaker.

Last year, the Service was instrumental in the Federal Government's response to Hurricanes Harvey, Irma, and Maria. The Coast Guard's hurricane response efforts were in addition to all of the everyday actions the Coast Guard does to protect our Nation.

One of which is drug interdiction. Last year, the Service had a record-breaking year for cocaine seizures, having stopped over 450,000 pounds of cocaine worth over $6 billion wholesale.

These efforts often go unsung, and we do not want that. We want the Coast Guard to be known for the valor of their servicemembers, because we know they take risks every day in defense of this Nation.

We need to understand the needs of the Service. Congress needs information to conduct proper oversight and to support programs with adequate funding. The lack of information on how the Service is meeting its statutory missions and how its assets are meeting or not meeting performance goals leaves Congress without much-needed information and could lead to inadequate budget requests, and more importantly, inadequate funding levels. We would like you to help us help you.

The Coast Guard already falls behind the other services. It is the only armed service subjected to nondefense discretionary budget requirements. So unlike the other services, this is not considered defense spending. This places the Coast Guard in competition with all nonmilitary discretionary spending despite the Coast Guard being a military service.

The lack of budget clarity that the nondefense discretionary budget has imposed on the Coast Guard has definitely impacted the Service in its ability to adequately and consistently fund its programs. As I have said before, without question, this is a risk to national security and should compel a more serious budget approach.

When the Service is active in time of war, it works as part of the Navy. But its everyday missions are critical to our national defense. I repeat, every day the Coast Guard's missions are critical to our Nation's defense, and the Coast Guard needs sufficient funding to acquire its assets and effectively do its job.

Admiral, this is your last hearing in front of this subcommittee. And on behalf of the ranking member, I am sure he will have something to say, too, but I just want to say thank you for your loyalty, steadfast service to this Nation. And we just enjoy being with you.

It is great to see the Coast Guard turn more in a Marine Corps direction. So as a marine to coastie, we just want to say thank you, and keep on weaponizing your boats.

MARAD is also with us today. The subcommittee shares jurisdiction over the Maritime Administration with the Armed Services Committee, having jurisdiction over the nonnational security aspects of the merchant marine. The subcommittee understands the critical role U.S. mariners have in supporting domestic shipping operations, as well as defense operations, including the Maritime Security Program and sealift.

MARAD has been an integral part of the subcommittee's Military to Mariner roundtables and discussions to assist active and retired military mariners to move into civilian mariner positions.

The subcommittee looks forward to working with MARAD on these important issues. I thank our witnesses for being here today, and I look forward to hearing your testimony.

I will now yield to Ranking Member Garamendi.

Mr. GARAMENDI. Mr. Chairman, thank you. Thank you for calling the hearing.

Once again, it is a welcome change to be able to talk about a funding increase rather than to try to figure out what to do with less money.

First off, Admiral Zukunft, thank you so very much for your service. You have been a pleasure to work with over this period of time. And the Coast Guard in the recent hurricanes performed admirably, way beyond the expectations—well, not beyond our expectations, but in a very, very difficult environment. And we know that has to do a lot with the command structure, and I know that you are at the top of that. So thank you so very much.

I also see that, Mr. Cantrell, you are also going to retire. I want to thank you for your service. And I am sure that all of the men and women in the Coast Guard appreciate what you have been able to accomplish representing them so well in your position as well as before this committee.

I guess we will look forward to Admiral Schultz and see how he bears up under our severe questioning. I am sure he will do very, very well with it, and I know that he has a significant person to follow.

Now, back to the budgets. It is not just about the numbers, but it is really how they all fit together. We have much better numbers to work with. That is not my concern.

Moreover, the real problem with this budget is, does it represent a change in which we can look forward to the future with wholesome, full budgets, or is it just an anomaly and we will go back to where we have been over the last several years, and that is trying to figure out how to do more with less.

Last year's budget, by example, requested the Coast Guard be cut by 14 percent. This year we are looking at an 11-percent increase. Well, we think that is a significant improvement. But what about the years in the future?

It is my hope that among the options, we are going to see increased budgets for the Coast Guard.

I note that the President is talking about his wonderful, beautiful border wall. I think he said $18 billion. Probably closer to twice that. But nonetheless, if you want to stop the drugs, as he apparently wants to do, better to spend that money on the Coast Guard where you have an extraordinary record.

Tariffs. Interestingly, the President unilaterally imposed tariffs last week on steel and aluminum, but in his budget he proposes not to use that newfound opportunity for the American steel industry to build ships, but rather for MARAD to go out and purchase a couple of used hulls and to somehow repurpose those for the academies. But what happened to the multiservice vessels that were authorized in the budget?

I am going to ask you, Admiral Buzby, about that. We think that we had a pretty good plan, and we are not interested in buying a bunch of foreign ships for the maritime academies, but rather to make them in America.

So what are we going to do? Well, we are going to find out where you are coming from today.

Additionally, the Title XI Maritime Loan Guarantee Program seems to have been zeroed out. So what does that mean for our ships that we need to build for the recycling program and for the Marine Highway Program?

All of this makes very little sense. We need to put together a budget that actually meets all of the needs and not cuts inappropriately.

So while the top line looks good, you get into the details and we have problems. We are going to get into that in detail.

I yield back.

Mr. HUNTER. I thank the ranking member.

And I just ask that we put on timers, too, because we never have anybody here at our hearings very often, and now we have quite a few people. We will stay on track.

Mr. DEFAZIO. Mr. Chairman, may I?

Mr. HUNTER. Absolutely.

The ranking member of the full committee is recognized.

Mr. DEFAZIO. Thanks, Mr. Chairman.

I would like to congratulate and thank Admiral Zukunft on his pending retirement. There was something about severe questions, and there don't need to be any today. I think we have come a long way in the last 4 years in terms of the Coast Guard's future looks better.

We are finally getting some of the recognition that all of us have been arguing for in terms of adequate funding for the resources you need. Instead of starting with a proposed cut, we have a proposed increase. Inadequate, but a start.

We are finally moving ahead with the first new icebreaker, but we have other things that are still underfunded. Although there is $135 million for constructing and renovating shoreside infrastructure, there is still a $1.6 billion backlog. We still have to look at a whole new fleet of helicopters, buoy tenders, heavy lifeboats.

There is an incredible list of things that we still need for the Coast Guard.

And when I think about this idiotic wall, I just think, boy, if we had the money that is being proposed for the wall that anybody can go over, under, or around, particularly on the oceans, wouldn't it be better to spend those billions of dollars on fully recapitalizing and meeting the needs of the Coast Guard and other real defenses to the integrity of the United States and to interdict smuggling.

Chief Cantrell, I also want to congratulate you. Thanks for your many, many years of service. And I don't know where you are going, I know he is going someplace warm, but I wish you well.

This was a new issue to me about OMB and the replacement ships with MARAD, and that causes me tremendous concern. And like the ranking member, I can't believe that we can't find American hulls. And even for that much money, I think we can build a couple of ships. I am not quite sure what the specifications are. But that seems like a fair amount of money. And if we are going to repurpose foreign hulls, I am going to be really not very happy about that. I think three, hopefully, at least, and maybe we can enlist the full committee chairman.

With that, I would yield back the balance of my time. Thank you.

Mr. HUNTER. I thank the gentleman.

Today we will hear testimony from Admiral Paul F. Zukunft, Commandant of the United States Coast Guard; Master Chief Steven W. Cantrell, Master Chief Petty Officer of the United States Coast Guard; Rear Admiral Mark H. Buzby, United States Navy, retired, Administrator, Maritime Administration; and the Honorable Michael A. Khouri, Acting Chairman, Federal Maritime Commission.

Admirable Zukunft, you are now recognized to give your statement.

Testimony of Admiral Paul F. Zukunft, Commandant, U.S. Coast Guard; Master Chief Steven W. Cantrell, Master Chief Petty Officer of the Coast Guard, U.S. Coast Guard; Rear Admiral Mark H. Buzby, U.S. Navy (Ret.), Administrator, Maritime Administration; and Hon. Michael A. Khouri, Acting Chairman, Federal Maritime Commission

Admiral ZUKUNFT. Good morning, Chairman Hunter, Ranking Member Garamendi, and ranking member of the full committee, Mr. DeFazio, distinguished members of this committee. I appreciate the opportunity to testify today and ask that my written statement be entered into the record.

Mr. HUNTER. Without objection.

Admiral ZUKUNFT. The Congress, and especially this subcommittee, has gone to great lengths to challenge the status quo, demand transparency regarding our needs, and serve as a strong voice for the men and women of your United States Coast Guard.

Examples, such as helping us establish the polar icebreaker integrated program office, your support in products like the unfunded priorities lists, typify what this committee has done for the United States Coast Guard. Your leadership and vision have helped us offset uncertainty in the budget cycle and have positioned us to receive the first meaningful funding increase since the Budget Control Act was passed in 2011.

And, yes, we are working with the administration to finalize our long-term major acquisition project and the latest Capital Investment Plan and unfunded priorities list. Many of these were submitted months ago. We continue to work with our Department and with the administration. We know you need those so you can help us in your endeavors to support the United States Coast Guard. You have my commitment that we will get these to you just as soon as possible, Chairman.

Following a series of devastating hurricanes, the Coast Guard launched one of the largest responses ever, culminating in the rescue of nearly 12,000 people. This was an all-hands-on-deck campaign, but it came at a cost. And I thank you and this Congress for the $835 million to address Coast Guard response costs and rebuild damaged infrastructure to modern standards.

And while so many endured these natural disasters, transnational criminal organizations continue to raise havoc in the Western Hemisphere. Last year, our campaign to protect the U.S. border, thousands of miles from our land border, netted over $6 billion worth of cocaine before it could reach our shores.

And equally important, we referred 606 smugglers to the Department of Justice for prosecution here in the United States where we prosecute 100 percent, less than 5 percent prosecution in their countries of origin, effectively advancing security and prosperity in our backyard.

Our cybersecurity program of record safeguards cyberspace and we secure our maritime critical infrastructure—ports, waterways, and commerce—that translates to $4.6 trillion in economic activity on an annual basis.

And we continue to be a sound investment. We earned our fifth consecutive clean financial audit opinion, the only armed service to do so, and our major acquisition programs continue to deliver assets that meet performance cost and schedule milestones.

Given our important work in the high latitudes, coupled with an ever-increasing assertive Russia and China, I am very pleased the President's budget includes $750 million for the polar icebreaker program as well. Our request for proposals, released just 2 weeks ago, has energized the U.S. industrial base and keeps us on track. We are as close as we have ever been in over 40 years to recapitalizing our Nation's polar icebreaking fleet, and we must keep this momentum going.

The budget also advances the recapitalization of our aged fleet of 35 inland construction and river tenders, some over 70 years old, in all with an average age of 56. Replacing these vessels with a modern, modest fleet of Waterway Commerce Cutters is a matter of economic and National Security Cutters as these ships are necessary to sustain our vital maritime transportation system, part of that infrastructure, if you will.

While this progress is meaningful and very encouraging, it is simply not enough. We need to continue building tomorrow's Coast Guard. Years of fiscal constraint under Budget Control Act caps have forced funding offsets and reduced our force structure. Going forward, we require 5 percent analyzed growth in our operations and maintenance account and a minimum of $2 billion in our acquisition account, a small ask for a service that provides great return on investment.

I have been honored to work alongside you these past 4 years. The more predictable the resource allocation process, the more ready our Coast Guard becomes. We are on the right track.

Thank you, Mr. Chairman, Ranking Member, and members of this committee, and I welcome your questions.

Mr. HUNTER. Thank you, Commandant.

Master Chief Petty Officer of the Coast Guard Cantrell, you are recognized.

Master Chief CANTRELL. Thank you, sir. And good morning, Chairman Hunter, Ranking Member Garamendi, and distinguished members of this committee. It is a privilege to appear before you today to represent the workforce of your United States Coast Guard.

As an Armed Force of the United States of America, as well as one of the Nation's lead maritime law enforcement and marine safety agencies, our men and women represent the finest coast guard in the world. These dedicated individuals are on the front lines every day carrying out major operations globally, as well as protecting our homeland at our maritime borders and inland waterways.

When disaster strikes, your Coast Guard responds with an unmatched humanitarian spirit. And I want to thank committee members and staff that have visited various Coast Guard units throughout a remarkable 2017.

While visiting our units, you heard directly from the selfless people who wear the uniform of one of our Nation's five armed services. I hope those visits provided you with ground truth about the effects your decisions have on our personnel, their families, and their careers.

I also want to thank you for your continued support of our recapitalization efforts, which, as you know, are making tremendous impact on an international level. Our new platforms are more capable, allowing our personnel to carry out their missions with the latest and most technologically advanced systems available.

We continue to have devoted Americans volunteer to serve and reenlist in our multimission service, and not just as individuals, but often as a family.

I, along with our senior leaders, remain deeply committed to providing our people with state-of-the-art platforms and world-class quality of life for them and their families as they serve this great country.

I would continue to ask that we not lose sight of the fact that we still have some work to do as many of our Coast Guard men and women continue to serve in cutters and at stations that are older and less capable.

But besides the challenges an aging fleet presents to our military workforce, in many places members and their families continue to face quality-of-life challenges. With many of our workforce serving at remote units far from bases or military installations, and in many cases high-cost coastal areas, access to affordable housing, adequate medical services, and suitable childcare will always be a challenge.

The sacrifices our members and their families make throughout a typical career are virtually impossible to quantify, and we simply cannot afford not to invest in them and their future.

I would like to take a moment to thank you for the language in the 2018 National Defense Authorization Act, which amended title 37, United States Code, section 476, allowing reimbursement of relicensing costs for our military spouses who are forced to move due to military relocation.

I also thank you for the support received in other important focus areas, such as professional credentialing and VOW [Veterans Opportunity to Work] to Hire Heroes Act requirements for our Coast Guard personnel.

We look at our new ships and aircraft that are the result of this committee's commitment to our Service. However, the efforts that we invest in our people are what breathes life into these awesome assets and allows our families to continue to support us.

These investments should be nonnegotiable in budget discussions. Investment and modernizing our aging accession points, along with our other shore infrastructure, should be a priority as it is just as crucial to Coast Guard mission success as new modern ships and aircraft. I ask that you ensure our fiscal year 2019 appropriation reflects your commitment to them.

I know you are keenly aware of these challenges, and I am grateful for the support from this committee as we continue to address each of them. While our recruiting efforts continue to be stronger than any time in the last decade, and we are meeting our goals, I reiterate that we must continue to invest smartly if we want to retain the talent that is lining up at our recruiting offices.

Mr. Chairman, members of the committee, on behalf of the servicemembers of your United States Coast Guard and their families, I thank you for your continued and very public support and for the opportunity to highlight some of our successes and challenges.

And as this will be my last opportunity to testify before you before I retire, let me personally say thank you. It has been a true honor to wear this uniform for nearly 35 years in service to our great country and such a remarkable opportunity to appear before you over the last 4 years to tell you about the needs of some pretty amazing Americans, United States coastguardsmen. Thank you.

Mr. HUNTER. Thank you, Master Chief. Admiral Buzby, you are recognized.

Admiral BUZBY. Good morning, Chairman Hunter, Ranking Member Garamendi, distinguished members of the subcommittee. I appreciate the opportunity to discuss the President's fiscal year 2019 budget priorities for the Maritime Administration.

The President's $696.4 million budget request for MARAD is focused on strengthening the U.S.-flag fleet, investing in the education and training of the next generation of merchant mariners, and supporting our Nation's commercial and national security objectives.

In 2014, U.S. maritime freight activity supported $4.6 trillion of economic activity, providing access to the world's markets and generating revenues equivalent to 26 percent of our GDP.

The U.S.-flag fleet in international trade, however, currently carries less than 2 percent of our annual foreign trade, and the U.S.-flag presence in international commercial trade is at a historic low of 81 ships. We must grow our share of overseas maritime commerce to remain economically viable globally.

As the U.S.-flagged deep-sea fleet dwindles and the U.S. becomes increasingly reliant on foreign-flagged shipping, we also negatively impact national security. The pool of U.S. merchant mariners who crew our ships in peacetime commerce are the same dedicated Americans we depend upon to answer the call to man our surge sealift fleet in times of national emergency.

The refusal of some foreign-flagged ships to deliver DoD cargo during Operation Desert Shield and Storm illustrates the danger of relying on another nation's shipping to move our military cargo in a crisis.

The Maritime Security Program, or MSP, provides direct annual stipends for 60 active, militarily useful, privately owned U.S.-flagged vessels and crews in international trade. This program ensures access to U.S.-flagged ships in international trade and, critically, the intermodal networks needed to move military equipment and supplies in national emergencies.

The President's budget requests $214 million, or $3.6 million for each ship enrolled in the MSP. This request is recognized as less than the fully authorized level for MSP. It reflects hard choices that had to be made as the administration pursues rebuilding DoD capabilities. The Department and MARAD strongly support MSP and recognize the critical contribution it plays in the Nation's security.

Funding from the Department of Defense in fiscal year 2019 will allow MARAD to maintain 46 surge sealift support and special mission vessels from the Ready Reserve Force in a 5-day readiness status and begin service life extensions for several ships as part of a comprehensive recapitalization program for this fleet, which has an average age of 43 years.

Survival of the U.S.-flag fleet and our ability to crew the surge sealift force, depend, deploy and sustain our military, requires an adequate pool of U.S. merchant mariners. Most U.S. Coast Guard credentialed officers who crew these ships graduate from the United States Merchant Marine Academy and the six State maritime academies.

The President's budget request includes $74.6 million for Kings Point, $70.6 million of it to support academy operations, and $4 million to fund capital building and infrastructure improvements for this 75-year-old institution.

Budget requests also includes $24.4 million to support State maritime academies, $2.4 million of which is to provide financial assistance to 75 cadets who agree to incur a Government service obligation and $22 million for the maintenance and repair of our aging fleet of federally owned training ships.

The administration is amending the President's budget request to include $300 million to replace two of the oldest training vessels to ensure the availability of safe and efficient vessels to meet critical mariner training needs now and in the future.

Finally, the budget requests $30 million for the MARAD ship disposal program, the majority of which will be used to continue the required decommissioning of Nuclear Ship *Savannah* while in protective storage.

These programs, supported by the President's fiscal year 2019 budget request, represent MARAD's top priorities. We will keep this subcommittee apprised of the progress of our program activities and initiatives throughout the coming year.

I appreciate the subcommittee's continuing support for maritime programs, and I look forward to working with you on strengthening our Nation's maritime transportation system and bolstering our fundamental, foundational Jones Act.

I request that my written statement be entered into the written record of this hearing. I am standing by for your questions, sir.

Mr. HUNTER. Thank you, Admiral.

I would like to recognize Mr. Khouri, but before I do, I want to recognize Rebecca Dye, sitting behind you, and Dan Maffei. Rebecca was also a committee staffer and a coastie. You have got the trifecta.

So welcome.

Mr. Khouri, you are recognized.

Mr. KHOURI. Thank you. Chairman Hunter, Ranking Member Garamendi, and members of the subcommittee, thank you for the invitation to appear today. With permission, I will summarize my written remarks and request that the written testimony and a copy of our fiscal year 2017 annual report be included in the record.

[The fiscal year 2017 annual report of the Federal Maritime Commission is available online at https://www.fmc.gov/assets/1/Page/Annual Report FY17.pdf.]

Mr. KHOURI [continuing]. I also recognize and thank Commissioner Dye and Commissioner Maffei for joining me today.

The FMC's mission is to ensure a competitive and reliable international ocean transportation supply system that supports the U.S. economy and protects the public from unfair and deceptive practices.

As the Commission monitors and regulates key sectors of the container shipping industry to assure competition and integrity for America's ocean supply chain, the Commission is meeting its mission, and our U.S. exporters, importers, and consumers are the ultimate beneficiaries.

We have seen significant change in the ocean transportation system over the last few years. The number of major global shipping companies decreased from 21 to 12. With new construction, global fleet capacity has increased to 5,200 containerships and 21 million TEUs [20-foot equivalent units] of capacity. This capacity increase outran global cargo demand, resulting in overcapacity in nearly all trade lanes.

Nine of the twelve major ocean carriers are members of three global vessel operation alliances. A reassuring data trend, however, shows us that even with the mergers and new carrier alliances, the individual ocean carriers continue to independently and vigorously compete on pricing and overall capacity decisions, providing evidence that healthy competition continues.

Industry analysts and shipper interests recognize that the alternative to well-regulated vessel alliances would be further consolidation in the industry with fewer ocean carriers and less service options.

Another positive development, carrier agreements that contain authority to discuss rates, have experienced a steady decline. Five such agreements terminated in the past few months.

The Commission has responded to these structural developments with new agreement negotiation practice and enhanced monitoring programs. We have insisted on narrower agreement authority and more clear, definite language.

For all agreements, our staff maintains a careful watch on industry trends, being vigilant for indications of anticompetitive behavior.

Marine terminals and port authorities have shown new interest in using cooperative agreements in their operations. Terminals are cooperating in new ways as they address the challenges of larger vessels unloading more containers at each port call and the related port infrastructure requirements and developing collective solutions to mitigate cargo bottlenecks.

On the regulatory front, the Commission continues to look for ways to aid compliance with statute and regulatory requirements for our stakeholders. Global supply chain operations benefit from less regulation through lower cost that pass through to our U.S. exporters, importers, and consumers.

Commissioner Dye recently presented the final report of the Supply Chain Innovation Team Initiative that highlighted the need for greater information visibility across the ocean freight delivery system.

In December 2016, the Coalition for Fair Port Practices filed a petition asking the FMC to begin a new rulemaking proceeding to address practices by marine terminals and ocean carriers related to demurrage, detention, and related fees. We received numerous comments and then held 2 days of public hearings in January. The Commission recently voted to institute a formal investigation to develop a full factual record.

Following her experience with the Supply Chain Innovation Team Initiative, Commissioner Dye agreed to serve as the fact-finding officer. The investigation report is scheduled for December of this year.

Regarding our budget, our requested level of funding for fiscal year 2019 is $27,490,000. The FMC is a small agency charged with a focused competition and commercial mission and a need for specialized staff, including a high percentage of economists and attorneys, career fields that tend to fall in the upper GS pay scales. The bulk of the Commission's budget, approximately 86 percent, is dedicated to these salaries and rents.

Thank you for your attention, and I will be pleased to answer any questions.

Mr. HUNTER. Thank you, Mr. Khouri.

I am going to now recognize Members for questions, starting with myself.

Commandant, as you look back over the last 4 years, just, if you would, highlight some things, lessons learned, saved rounds, and the biggest challenges that your successor is going to have to face, from your point of view.

Admiral ZUKUNFT. Thank you, Chairman.

So as I look back 4 years ago, we were very focused on two major acquisitions, and that is security cutter, but more importantly, the Offshore Patrol Cutter. But we didn't really look outside the scope of those major acquisition programs.

We started applying a strategy driving a budget approach to the United States Coast Guard, rather than coming to you asking you what we need, but explain why we need it, but how that fits within the national security framework.

As we look at where the Department of Defense is going within the national military strategy, but where are they not going? You don't see a national military strategy that addresses the Western Hemisphere, nor do you see one that addresses the Arctic. As an armed service, we can close those gaps, and then with the unique authorities that we have, we can play in all of those areas.

So with the support of this committee, we are now modernizing our entire fleet of Coast Guard cutters: the National Security Cutters, Offshore Patrol Cutters, the Fast Response Cutters, polar ice-breakers, and now our Waterway Commerce Cutters.

We have created a cyber program of record. We have a Cyber Protection Team up and running. Next year, we will have a cyber curriculum at our United States Coast Guard Academy.

Daily we sail into harm's way. For the last 2 years, we have had record removals of cocaine. I have been able to meet with the Presidents of every country, except Nicaragua, in Central America, multiple times with President Santos, to put pressure on them to stop the movement of drugs coming through their country destined for the United States.

In 3 weeks, the United States Coast Guard, Mexican Navy, Colombian Navy, we will be patrolling together for the first time ever in the eastern Pacific.

But as we sail into harm's way, and as we saw during last year's hurricane season, in the last 4 years we have not lost one Coast Guard life in the conduct of conducting our hazardous operations. And to that, I owe great credit to our officers in charge, our commanding officers, to balance risk and at the same time carry out some of these, quite honestly, very heroic missions.

And as we are modernizing, we have not been doing it at the expense of force structure. We are growing the force and we are growing the fleet.

There is no better time to join the Coast Guard than right now, and we are seeing that at our recruiting stations, the people that not only want to join the Coast Guard, they want to stay in the Coast Guard.

So what concerns me as I look out beyond 2019? Uncertainty. We have got two tremendous at-bats, as I like to call it, in 2018 and 2019. But there are many more at-bats.

And was mentioned, is this an anomaly? And so what we need is going to be predictable funding as we look in the out-years going forward.

That will be the most critical aspect, is to convey the value proposition of the United States Coast Guard, demonstrate that we can be accountable for these dollars, as we have had the last 5 years with clean financial audit opinions, with responsible acquisition programs that deliver on-time, on-budget programs that easily meet and many times these new assets exceed our operational requirements, as we have seen with the National Security Cutter.

So challenges ahead, but, again, with the tremendous support, and taking away some of those huge uncertainties, including what could have been a 14-percent reduction to our funding base. We would not have prevailed last year during Hurricanes Harvey, Irma, and Maria if we did not have that fiscal support.

So support here, challenges ahead, but we are clearly on the right trajectory. And really, since 2011, we are finally seeing an upward turn in our appropriation.

Thank you.

Mr. HUNTER. Thank you, Commandant.

Master Chief, same question, what do you see as the challenges for your successor? And I thought you guys did left-seat, right-seat carryover to the next Commandant, but you don't. The next Commandant brings in his own senior enlisted person, right?

Master Chief CANTRELL. That is right.

Mr. HUNTER. So what do you think, Master Chief?

Master Chief CANTRELL. Well, and I would like to talk about a couple of successes and accomplishments. First, we got—well, he has left—we got Congressman DeFazio out in a motor lifeboat out in the middle of a Columbia River bar last year and got him wet. So that was an opportunity for him to see the real good work that our folks did.

But over the last couple years, we have had a couple of housing areas that have opened up. We have made some modest investments in some of our older housing. We have been able to get our childcare reimbursement program enhanced so that our folks are getting their reimbursements a lot quicker. And with the help of this committee, particularly over the last 3 years, of being able to protect our Basic Allowance for Housing, particularly for dual military couples, and I appreciate that.

And we have really put our focus on leadership development from the time that they leave boot camp all the way through to where they are master chiefs, so that we are really training our reliefs and our reliefs' reliefs in the best possible way.

And one that we are really proud of is, as an armed service, 4 years ago, our recruits were not firing live ammunition at boot camp. They were using a simulator. And we were able to turn that around and get our folks and find resources to get them to where they are actually going to a range and shooting live ammunition and getting qualified before they go out. And as a leading maritime law enforcement agency, that was just really, really important, and we have been successful there.

For my relief, still got work to do. I think we still need some more childcare centers. And while we do have money for this year and next year to get a new child development center, we need to continue to work that, particularly in some of our remote areas that don't have access to DoD facilities.

A continued investment in our accession point. Our basic training facility in Cape May is old, it is in need of repair. There are some plans over the next 10 years to rehab that, and we just need to keep the pressure there.

And professional credentialing and licensing, which was required by the NDAA in 2015. We need to get after that and give that to our folks as they transition out of our Service and give them the things that they need.

And those are the things that I will pass along to my relief later this month, hopefully.

Mr. HUNTER. Thank you, Master Chief.

And, Commandant, again, the last 4 years, I realize this is my sixth budget hearing on this, the last 4 years I think was a sea change for the Coast Guard in the direction that you are going, and it has been an honor to be a part of it.

Mr. Garamendi.

Mr. GARAMENDI. Thank you, Mr. Chairman.

We have spent the last 4 years trying to figure out how we are going to get a heavy icebreaker. We are well on track to do that.

Admiral, could you bring us up to date on where we are with the heavy icebreaker and, specifically, the funding issues for the first, second, third?

Admiral ZUKUNFT. Thank you, Ranking Member. And as you know, as we look, it is not finalized yet, but certainly the 2019 budget, very optimistic, with an appropriation that gets us an ice-breaker with DHS funding.

As we all know, there is language in the NDAA, the National Defense Authorization Act, that puts the onus on the Department of Defense to provide one icebreaker.

So there is still more work to be done in terms of how that funding is sequenced. We need more than one icebreaker. The program of record, it is six. And then going forward, we need to look at block buying these.

And there is language in the request for proposals for the five potential contractors to submit that as well so we can make an informed decision, because this is the highest risk mission that we have right now. I have nothing else I can send down to Antarctica in my inventory right now if our only heavy icebreaker should become beset in ice.

So that is the additional work that needs to happen going forward with our polar icebreaker program.

Mr. GARAMENDI. The budget augmentation has some $700 million. My personal opinion of that, that is for the second and third icebreakers and we ought not relieve the Department of Defense from funding the first icebreaker.

That is where I am coming from on it. We will see how all that works out into the future as we put together the new NDAA and the appropriation, the omnibus now underway, then the 2019 appropriations.

I toss that out and say that is the bottom line, and hopefully that will be the way it is, so that that $750 million doesn't relieve the Department of Defense from its current obligation to fund the first icebreaker.

Going forward, we have questions, really I want to get to MARAD and Admiral Buzby. And this has to do with the National Security Multimission Vessel. Once again, the addendum to the budget indicates that there is $300 million to buy some laid-up, used, rusty foreign-built hulls, rather than the new National Security Multimission Vessel.

Let's get into it, Admiral Buzby. Where are we with this? And are you really serious as a representative of the administration to toss out all the work that has been done and the design? And by the way, where is the design of this new multimission vessel?

Admiral BUZBY. Thank you for the question, sir.

First, the design is alive and well. It is fairly mature and could be built.

Mr. GARAMENDI. Fairly mature is what I would call squishy words. Exactly where is the design? Is it completed?

Admiral BUZBY. It is not ready to build from. It would need to go one further step into detail design and advanced procurement.

That would be the next step, which would——

Mr. GARAMENDI. So it is good to go to the next step. However, the administration is taking us to a bunch of rusty buckets.

Admiral BUZBY. Well, sir, the reason, I think, that the request was made the way it was, was due to the timeframe. Given the funds that are potentially in the fiscal year 2018 budget, which we don't know for sure yet, given the fact that we are bumping up against the stops with the *Empire State*, having to get something in place for her in 2019, the nearest path, now that we are kind of getting in against the stops, was to go on the open market to find, not necessarily a laid-up ship, a ship that is available.

And it could still be a U.S.-built ship. We are going to cast a very wide net to see what is available to then upgrade.

Mr. GARAMENDI. So if we follow that, what happens to the whole concept of a multimission vessel?

Admiral BUZBY. I think it is still alive. Mr. GARAMENDI. I think it just sunk.

Admiral BUZBY. The business case is still there that a new ship should be procured. That still exists.

If money were to materialize and we could get started on a ship sooner, a new ship, we could build a new ship. We have the capability to do that. But the time is what is working against us right now.

Mr. GARAMENDI. Let me just be very clear: No way, no how are we going to allow you to go out and buy a bunch of rusty old hulks rather than continue with what we have been working on for the last several years, and that is the concept of a multimission vessel. So keep that in mind.

Admiral BUZBY. Yes, sir.

Mr. GARAMENDI. There is $300 million that is somewhere out there. And I will do everything I can, and I think the committee may very well be supporting it, at least the ranking member indicated he did.

Admiral BUZBY. I look forward to working with you on that, sir. Mr. GARAMENDI. My time has expired, and we are going to do the best we can to stay to our 5 minutes.

Thank you, Mr. Chairman. I yield back.

Mr. HUNTER. I thank the gentleman.

Mr. Mast is recognized.

Mr. MAST. Thank you, Chairman.

I will just start by saying, Master Chief, I think you are at about 2 months and a wakeup from what we talked about. Congratulations. Thank you for your service to this country. It is incredible.

It has been an honor to be able to question you numerous times out here. And I am just very proud to have you join us numerous times on these panels here. So thank you for your service to this country.

Commandant, I have just got a couple of quick questions for you, one local, one more broad about airframes. But locally, at my level, you guys have been doing a great job doing a navigation feasibility study in my area on our big waterways. If you could just have your folks get me a completion date on that, it doesn't have to be in this hearing, but get me a completion date on that, I would appreciate that greatly. I just thought I would take the opportunity to ask for that.

Looking more at the airframes, what I wanted to ask was about this. And I understand that the Coast Guard is looking at a lot of different courses of action to keep the fleet of MH–60Ts up, operational, your workhorse for search and rescue. Love having them in the area. You know, I am a boater that gets out in Fort Pierce and Fort Lauderdale and Miami, so I like having you guys out there.

But if you are going to keep those flying until the 2030s, I am also told that those average flight hours on those airframe are about 15,000 hours, so it is a pretty good amount of time. And, of course, we all realize on this committee how the Coast Guard is really constrained in terms of resources.

And so in that, I was wondering if you could answer, have you guys been looking at the purchasing of new cabins or airframes that would basically zero-time those aircraft? Is that something that you all look at?

Admiral ZUKUNFT. Thank you, Congressman, and we will get those waterway studies to you as well.

So with our two helicopters, the H–60 Sikorsky, which is our long-range helicopter, we are actually taking what we call sunset airframes from the other armed services, and we completely renew these, re-engine them. And what we end up with is a like-new H–60 airframe. We reset the hours on it. And we are acquiring these at half the cost of a new airframe.

Our H–65 Dolphin model is now in its fifth spiral development, if you will. They have been re-engined. It is a composite airframe. And when I first deployed with these helicopters on my ships back in the late 1980s, when they first hit the fleet, they actually have a much greater reliability today than they did when the first generation of this airframe came out. And much of this is being done through self-help.

As we look at our 20-year investment plan, starting in about the year 2033 is when we start recapitalizing our short-range airframes. And then overlaid with that is going to be our long-range H–60 airframes.

And the key decision point at that point will be, what do we have in our unmanned aerial inventory? We are also working on manned aerial systems to include land-based. And then, do we look at a one-airframe model or do we look at a two-airframe model, recognizing there are great economies of scale, training, and the like, inventory, parts management, if you have one airframe?

And then looking across the Department of Homeland Security, especially within Customs and Border Protection, to look at commonality across the Service, but we are also part of a joint service, to make sure that we have interoperability there as well.

I feel quite comfortable with the timeline that we have set right now, just given the reliability and the work that we are doing organically within the Coast Guard to maintain these airframes.

Mr. MAST. So this helps you stretch the dollar, and you could easily say this is much more than a Band-Aid, this is good long term.

Admiral ZUKUNFT. Near term and long term. And these are more than Band-Aids. We are flying these 800 hours a year. The reliability rate, we had over 90 airframes down in Harvey, Irma, and Maria, not one of those airframes missed a mission. I think it is a testimony to the wrench turners that often don't get the due credit, the maintainers, if you will, of this squadron of aircraft.

Mr. MAST. Thank you. Thank you, Commandant. And, Chairman, I yield back.

Mr. HUNTER. I thank the gentleman. Ms. Plaskett.

Ms. PLASKETT. Thank you, Mr. Chairman. And thank you, gentlemen, for being here.

Admiral, thank you so much for the work that you all have been doing in the Virgin Islands and the Caribbean. And you and I have had several encounters in both good and bad times, and I am grateful for the support that you have given to the people of the Virgin Islands and Puerto Rico as well.

Following the passage of the Bipartisan Budget Act, for the fiscal year 2018, the administration delivered a wish list of how it would want to have that funding spending.

I guess my question to you would be, is that reflective of where the Coast Guard most needs additional funds for the rest of fiscal year 2018?

Admiral ZUKUNFT. Yes, ma'am. And what I am most encouraged by is the supplemental relief that we are seeing, especially from Hurricanes Irma and Maria.

As you well know, the U.S. Virgin Islands and Puerto Rico, these are Coast Guard communities as well. And so when you see the Coast Guard is still there, as these communities are still trying to get back on their feet, these are Coast Guard families in that very same predicament as well.

So what this supplemental relief does is, rather than build a house out of straw, let's build one out of brick and mortar. In other words, let's harden these facilities so they could withstand at least a category 3 hurricane.

We made a similar investment out of Great Inagua with our hurricane hangar out there. It experienced 155-knot winds during Hurricane Irma. It sustained no damage whatsoever.

So we need to use these dollars and invest smartly because we are in the U.S. Virgin Islands and Puerto Rico for the long haul. These are Coast Guard communities as well, and we are proud to be part of those communities.

Ms. PLASKETT. OK. I also wanted to ask you, the majority of the illegal contraband is being interdicted in the eastern Pacific, but you know there is a significant drug smuggling threat that reaches the U.S. Virgin Islands and Puerto Rico from South America.

How will the additional National Security Cutters and the future Offshore Patrol Cutters defend against this threat? And do you believe that it is the sufficient amount and the appropriate number? Would you want more vessels to be able to take on this threat in that area?

Admiral ZUKUNFT. So numbers do matter. And the National Security Cutter has clearly been a game-changer. We are seizing nearly a ton of cocaine a day, and that is between the Caribbean, the eastern Caribbean, and in the eastern Pacific today.

What has changed in the last year is we have seen a significant decrease in the number of Cuban migrants attempting to cross the Florida Straits. We have a fleet of Fast Response Cutters home-ported in San Juan, and we have been able to repurpose those ships and apply a lot more pressure on the approaches to Puerto Rico, to the Dominican Republic, and to the U.S. Virgin Islands for shipments that are leaving the Guajira Peninsula on a direct run, because the flow rate coming into Puerto Rico right now has gone up over 30 percent.

Ms. PLASKETT. It has grown over 30 percent?

Admiral ZUKUNFT. It has grown over 30 percent over the last 2 years. It is not staying in Puerto Rico. It is being transshipped elsewhere, to include the U.S. Virgin Islands and to include the continental United States as well.

So we are repurposing resources. All this based on intelligence. Ms. PLASKETT. And is it the sufficient amount that you need to be able to address this increasing threat in the area?

Admiral ZUKUNFT. The U.S. will never get in front of this problem by itself, which is why we need allies. The interagency is contributing as much. And as many ships as we have put out there, what we have soon run out of is surveillance aircraft, which is why we are looking and experimenting with land-based UAS. We have done a couple of prototypes out of Puerto Rico as well.

We need to look at state-of-the-art sensors to put into these unmanned aerial systems. But the intelligence is good. We do not have enough ships or aircraft to be fully effective in this mission.

Ms. PLASKETT. Thank you.

Administrator Buzby, thank you so much, sir, for your service.

And I wanted to ask you, with regard to the Maritime Administration, can you tell us what, if any, activities or support that your agency and your group, particularly your Ready Reserve Forces and your own fleet, were able to provide during the hurricane season in the Caribbean?

Admiral BUZBY. Yes. Thank you, ma'am.

We were very proud to play a significant role, clearly not as significant as the Coast Guard who were heavily involved. We had three of our school ships and one of our Ready Reserve Force ships active, participating in hurricane relief for all three hurricanes. We were active in Texas, we were active in the gulf coast, in Florida, and most heavily in Puerto Rico and the Virgin Islands.

Ms. PLASKETT. Because I know that one of the issues that we had on the islands was that the inability to have cargo, particularly food relief and others, come in. I know that most of your support is not related to those things, but related to military equipment.

But could your fleet also be used to support the additional food and other supplies and equipment that were needed in places like Puerto Rico and elsewhere?

Admiral BUZBY. Actually our Jones Act fleet of commercial ships, we had 24 ships that were providing support to Puerto Rico throughout the thing. So there was plenty of capacity.

But to directly answer your question, yes, there is additional capacity that could be brought into play with our Reserve ships to augment that should it be necessary.

Ms. PLASKETT. OK. Thank you.

Thank you very much. I yield back.

Mr. MAST [presiding]. The Chair is going to recognize the gentleman from Louisiana, Mr. Graves.

Mr. GRAVES OF LOUISIANA. Thank you, Mr. Chairman.

First of all, Admiral, I want to thank you very much for your service to the Coast Guard and to the Nation. I think we first started working together about 8 years ago when we were both in different capacities. And I remember one of us getting fairly excited and maybe yelling a good bit and things like that across the table at one another. And thankfully, your temper has subsided since then.

I am kidding. I might have been the one doing that.

No, but, again, I just want to thank you very much for your service to our Nation. I also know that your wife and family have sacrificed much for your service in the time away, and I do want to thank you.

I know that Chairman Hunter talked to you a little bit about some of your lessons learned and advice on the way out, and we are going to keep your cell phone number and keep bothering you as things progress to continue to get your advice on how we best handle maritime policy, maritime security, and many other challenges that our Nation is facing.

But your service to the Coast Guard was one, as we have discussed at the last few hearings, one of where the Coast Guard has evolved into sort of this Swiss army knife in this post-9/11 and evolving-threat era that we are in now, where the Coast Guard's presence on the water has just continued to evolve as the threats have. And you all are an extremely agile, multimission agency. And, again, your leadership during those challenging times has very much been appreciated. So thank you very much.

Admiral ZUKUNFT. Thank you.

Mr. GRAVES OF LOUISIANA. Chief, same to you. I always appreciate your absolute, unconditional support for the coasties out there and advocating on their behalf. You have been a great voice, a very passionate voice for the men and women of the Coast Guard. And have appreciated very much your fighting for the people that are on the ground, that are the people that are literally protecting our Nation on a daily basis. And I want to thank you for your service as well.

Admiral Buzby, before I jump back to Coast Guard, I want to jump over to you just for a minute. You have some pretty strong words for the importance of a domestic fleet in your testimony. You talked a little bit about the important role that Jones Act vessels play, U.S.-flagged vessels play in terms of ensuring that we have the capability to deliver our military men and women, our equipment around the globe.

And you also talked about some of the challenges associated with our increasing reliance upon foreign-flagged vessels. I believe you noted in your testimony that there were 177—give me a little lee-way there—foreign vessels that were used. And I believe you noted that in at least 13 cases that those foreign vessels had apprehension, hesitation, or refused to go into some of these areas where there was tension or conflict. And that, obviously, is exactly where we need our military men and women and exactly where we need our equipment.

Do you care to expand or comment on that and just sort of your vision on how we ensure those types of situations don't happen in the future?

Admiral BUZBY. Thank you for that question, sir.

It really is critical to my way of thinking, and I think to most of the military, and certainly to General McDew of USTRANSCOM [U.S. Transportation Command], that we have reliable sealift for this country. And that has to come from a combination of Government-owned ships that we hold in reserve in our sealift fleet, but more importantly, from the U.S.-flag fleet. Those are the ships that we can rely upon.

And some of the cargo that we have to move is critical to what-ever mission we have to do. And to entrust that to a foreign-flag ship with an unknown crew, and as we have seen historically, some of those declined to do that. They declined to carry that to where we needed to go. And that is simply unacceptable. We depend on our sealift forces to move our military overseas, period.

Mr. GRAVES OF LOUISIANA. Thank you. And I do appreciate you providing us that feedback or guidance. And I just want to ask if we can continue to work perhaps offline a bit on steps we need to take here to ensure that we don't run into those reliability problems in the future. But I think it does, just to reinforce as this committee has on numerous occasions, the importance of having a robust U.S.-flag fleet that can deliver for us.

Admiral, coming back to you, Admiral Z, the icebreaker is something we have talked about a good bit over the last few years. We have had a number of meetings discussing that. I always get this mixed up—the *Polar Star* is the one that is still floating, right? All right.

So the *Polar Star*, we are in a situation now where the service life of the *Polar Star* is already, while you all did a phenomenal job with duct tape and bubble gum keeping that thing floating, the service life on it is somewhat tenuous, I think it is safe to say.

Yet, as I recall, the Coast Guard has indicated that a $2 billion annual investment in a heavy is really what we need to get us to the place we need to be, to have the capabilities we need, and also to have a seamless transition from the *Polar Star* into a new heavy. I think it is fantastic that the budget request does have, was it $725 million in there, for a heavy for fiscal year 2019.

Can you just comment on sort of how those things transition? Because if the $2 billion figure was our target, if the service life is, what is it, 2023, I think, on the *Polar Star*, it seems like we are going to have a problem, if we are not fully investing in the program annually.

Admiral ZUKUNFT. Yeah, I am concerned, as I look at the 5-year budget bill, when I am asking for a $2 billion floor, and we are seeing our acquisition budget continuing to be funded below that floor.

And icebreaking is clearly the biggest risk. There is a $15 million annualized appropriation to keep the *Polar Star* in service, so there is a smooth hand-off between the *Polar Star* and the next heavy icebreaker.

That gets us right back to the status quo, though, and we are still only a Nation of one heavy icebreaker, which means we need to continue to build out that program of record. We need to provide predictability for our shipbuilders as well. Our shipyards have not built a ship of those scantlings, of that design, laying up that much steel since these ships were built over 40 years ago.

It is an investment in our industrial base at the same time. And this will be built in the United States and with United States steel as well.

So of all our appropriations, this is the one at greatest risk, and it does concern me. And there is going to be tension as we look at, how do we fund other priorities within the Department of Homeland Security?

And a concern that I will pass on to my relief is you may enter another cycle of flatline budgets at a point in time where our needs are continuing to grow, particularly in this domain.

Mr. GRAVES OF LOUISIANA. OK. Bottom line is, Mr. Chairman, I just want to highlight, I think we do have some problems with this current schedule in that the service life on the *Polar Star*, we are looking at 2023. Am I remembering that correctly?

Admiral ZUKUNFT. 2023 is when we anticipate taking delivery of the first heavy icebreaker. We want to sustain the *Polar Star* for 2 years beyond that to make sure as we go through sea trials we are fully mission ready.

Mr. GRAVES OF LOUISIANA. OK. So we have got major concerns with the *Polar Star* still floating by, I think, 2025 then.

Admiral ZUKUNFT. That is right.

Mr. GRAVES OF LOUISIANA. And actually having the ability to replace it with a new heavy. And then, of course, we have you, in discussions in another formal setting, the administration has indicated the need also to have medium capacity. And so that is going to need to come on. And if all we are doing is $700 million, then we are not able to even deliver the heavy, much less coming behind on the medium, which is really important.

If I can change gears real quick. Patrol Forces Southwest Asia has six 110s right now. We certainly want to make sure that the men and women that are out there helping to support our military in that region, that they have the equipment they need, have the resources they need.

I am a little partial to the FRC [Fast Response Cutter], as you know, and I just am curious if you could talk at all about, perhaps, the need or timelines on replacing the 110s with FRC out there.

Admiral ZUKUNFT. Yes, Congressman. I have had a number of high-level engagements. I have written when we want to restore readiness, a letter that went through the Secretary of Homeland Security that I addressed to the Secretary of Defense, to the Chairman of the Joint Chiefs.

I have met with General Votel, CENTCOM [U.S. Central Command], I have met with Vice Admiral Aquilino, he was commander of the 5th Fleet in Bahrain, with the Secretary of the Navy, and with the CNO, and that we have a hot product line that are building these ships on time, on schedule, because the 110s that are there will time-out in 2022.

The PCs that the Navy operates, they will start timing out a year later. So there is an opportunity, using DoD funding, to keep this product line going, to reconstitute the fleet that is over there right now. And we need a decision by 2022 so we can make those informed decisions to keep that product line going, to roll out that next fleet of ships.

So, Congressman, that is where it stands right now. And it does have the highest level of attention from the Secretary of the Navy on down, to include the combatant commander in charge of that theater. The return on investment is huge. The mission is not going away. The demand signal remains high. And so that is where we stand right now.

But 2020 is where the shipbuilder needs to be informed that this order is coming in so they can keep that product line going.

Mr. GRAVES OF LOUISIANA. Great. Thank you.

Mr. Chairman, there is this weird clock thing in front of me, I am not sure what that is, but I want to yield back.

Mr. HUNTER [presiding]. The ranking member and I stayed under our time so that we could give everybody else a chance. We are just happy you are here.

Mr. GRAVES OF LOUISIANA. Well, thank you very much. Mr. HUNTER. Mr. Lowenthal is recognized.

Dr. LOWENTHAL. It is a pleasure to join the Garret Graves show. Admiral Buzby, I want to ask what to me has become kind of a conundrum, and let me lay it out. It has to do with the President's budget request for MARAD's Title XI, the loan guarantee program. You know, the program has been in operation for over 80 years. It guarantees financing for the construction of American-made U.S. flagged vessels. It is a longstanding public-private partnership. It has helped to build the U.S.-flagged Ready Reserve Fleet, supported American shipyards.

These loans were crucial to help finance the construction of the world's first LNG-powered containership, which was built here in America.

President Trump talks about reviving American manufacturing and just last week announced tariffs under a rationale that certain industries are vital for our national security. I thought for years we have considered shipbuilding one of those vital industries.

My first question is, and let me ask the whole thing, is that still the case? And help me to understand the logic that if this program has an important role, which you say, and the President talks about American manufacturing, but the budget calls for no new authority to guarantee loans. That seems to me to be a contradiction.

Can you respond to that, please?

Admiral BUZBY. Yes, sir, happy to.

We believe that it is an important program also. This really comes down to a funding issue in a very difficult budget year. It is a good program. It has built over 1,900 vessels for this country.

It has done some good things. It has had some defaults in the recent past, which have been difficult to get through, but it has been a general benefit to the Nation.

Dr. LOWENTHAL. So you are very supportive of the program and the importance of the program?

Admiral BUZBY. I think it has made great benefit to our country.

Dr. LOWENTHAL. Thank you.

Acting Chairman Khouri, I have a question. Last year the subcommittee passed legislation to amend the Shipping Act to increase FMC's authority to oversee and enforce antitrust requirements that would prevent the ongoing consolidation among American—among foreign-flagged containers, not American, foreign-flagged containers, container carriers, from disadvantaging the U.S. maritime and port service industries.

And I have been told we expect Congress to clear this legislation later this year. So my question, when final agreement is reached, I think, on the Coast Guard Authorization Act—there are two questions.

One, what is the status of consolidation in the container trades?

Are there now fewer carriers than there were in 2017?

And is there sufficient funding in the administration's request for the FMC to ensure that the Commission has the personnel and resources necessary to conduct very effective oversight during this very kind of dynamic period of consolidation that is taking place?

If you can tell, do you have the resources? And what is the latest status on consolidation of the U.S. container trades?

Mr. KHOURI. Thank you for the question. Let me answer the second one first, if I may.

We do have the resources to continue our monitoring of the ocean industry, ocean container industry. We do have in our budget money for additional people, especially in our economist group. We just received notice that we are going to lose two economists via retirement. That area is sort of the heart and soul of everything that we do.

So we are actively getting ready. I think tomorrow we will have new notices for recruiting economists. These are somewhat specialized transportation economists. But I do feel confident that we can continue that mission and do it effectively.

On the status, as I said in my testimony, the number of major carriers has gone from 21 down to 12. There are a total of 36 different carriers that currently serve U.S. trades. Now, obviously, the smaller carriers are not up to a scale as a Maersk or Mediterranean

Shipping. However, bear in mind to some extent, when you have a commoditized marketplace, that the smaller carriers do effectively, from a pricing standpoint, set a lid on how all of the other carriers react.

We watch like hawks, I can promise you, the pricing activities and all of this pricing and decision—excuse me, capacity decisions are regularly filed with the Commission. We follow that closely. We see where different carriers are making pricing decisions this way and that. They are making their short-term decisions on putting ships in or out of a trade lane.

And then there are long-term decisions, similar to the Coast Guard, when you build a ship you have got it for 30 years, that the long-term decisions that various carriers are making in terms of new-builds continues to show diversity.

So this is all evidence to us that you still have a very competitive marketplace.

Dr. LOWENTHAL. OK.

Mr. KHOURI. So I think, as we speak today, the Herfindahl-Hirschman Index, not to get technical, but this is the same type of analysis the Department of Justice uses to look at industries, et cetera, that our main East-West trades, Asia to U.S., and the Atlantic trades are still in a very comfortable, what is called safe harbor area, under HHI.

So we feel confident that it still has a very competitive marketplace overall to protect our exporters, our importers, our consumers, and the various service providers that you mentioned in terms of the tug companies. So we do feel confident. We would like to have full funding so that we can go ahead and hire some of these extra folks into the 2018 and 2019 budget periods.

So that is where we are at this point. Thank you, sir, for the question.

Dr. LOWENTHAL. Thank you. And I yield back.

Mr. HUNTER. I thank the gentleman.

We are going to start another round of the hearing by recognizing myself. This is more what it looks like usually here.

FMC questions really quick. We all raised concerns last year. And, by the way, we have a bipartisan FMC bill, that is a Senate bill as well, so it is bicameral, bipartisan, to help do the things that we talked about last year. I think you got yelled at by just about everybody up here.

So in regards to that, we talked last year about you agreeing to allow a group of international car carriers to collectively negotiate with U.S. tugboat service providers. In light of the concerns that we have raised, is the Commission now more closely overseeing negotiations between ocean common carriers and the U.S. domestic providers?

Mr. KHOURI. Thank you for the question. The very short answer is yes. And just perhaps another sentence or two.

It has been interesting that some of the alliances have visibly backed away from even pursuing some of that activity. So I think it was heard by everyone in the industry where the Congress stands on the issue, and I feel comfortable that it is in good shape.

And I think the other thing that is in the, if I may, in the pending bill, I mentioned in my testimony that five different price discussion—what we call price discussion agreements, agreements that allow for discussion of pricing, I just note, I am not going to say because of, but I note that they have filed agreement terminations at the Commission. So that is a very positive trend of eliminating price discussion through the trades.

Mr. HUNTER. Mr. Khouri, then, you have seen the legislation that we have put forward, right?

Mr. KHOURI. Oh, yes.

Mr. HUNTER. Do you think that gives you more tools in your tool-box?

Mr. KHOURI. Yes, I do. Yes.

Mr. HUNTER. OK.

Admiral Buzby, I would ask you more questions, but in the NDAA and our other committee, we are both on the Subcommittee on Seapower and Projection Forces, we are both, John and I, we are both on the Armed Services Committee. You get authorized by the NDAA for MSP, it gets paid out of DOT, and the difference, the delta between what we authorized, the $300 million and your $216 million that you are putting towards MSP with the old $3.5 million number, instead of the $5 million per ship, we are going to work that.

I am sure we are going to find the money in the Armed Services Committee, because even though DOT pays for that, the difference needs to be made up in the NDAA, from what I understand. So we will work that going forward.

I guess the one question I have is, do you expect if the number stays at $3.5 million, you have 59 MSP ships now, right?

Admiral BUZBY. Sixty.

Mr. HUNTER. You have 60 now?

Admiral BUZBY. We are up to 60.

Mr. HUNTER. What do you expect that to go to? Because when we made this change it was for a reason. It was because ships were dropping out, right?

Admiral BUZBY. Right. We work very closely with our operators. They are very supportive of our program. I do not anticipate anybody that will drop immediately. We will be working very closely with them, however, if it has to stay at that lower number. It is a concern, no doubt. They have voiced their concerns. We are concerned about it. We would have liked to have absolutely funded it to the $300 million level, but just budget-wise we were restrained.

Mr. HUNTER. But also it looks like you have got money in to build two merchant mariner training ships—not to build, to buy.

Admiral BUZBY. Right.

Mr. HUNTER. Right? And those would not be U.S. built?

Admiral BUZBY. They could be.

Mr. HUNTER. They could be U.S. built, but the numbers you have I don't think are.

Admiral BUZBY. Well, if we have to go out and get used ships, the estimate is that to purchase a used ship and then convert it, it would be about $200 million or so to do that for a used ship to then modify it.

Mr. HUNTER. And right now you have two in there, right, you have the money for two?

Admiral BUZBY. Right now the language calls for two for $300 million.

Mr. HUNTER. OK. But you could theoretically build one.

Admiral BUZBY. We could build—that same number, that 300 number——

Mr. HUNTER. And put the money back to MSP.

Admiral BUZBY. It is theoretical, but, yes, sir.

Mr. HUNTER. Those are all possibilities.

Admiral BUZBY. It is a possibility. And we look forward to working with Congress on this.

Mr. HUNTER. Thank you.

Mr. Garamendi.

Mr. GARAMENDI. Just to follow up on some questions the chairman raised.

How many ships would be in the MSP fleet if the 29-percent cut were to take place?

Admiral BUZBY. In other words, if we were to fund it to the $3.5 million, we believe that all 60 ships would remain, sir, at this point at least.

Mr. GARAMENDI. So you intend to negotiate down the subsidy per ship?

Admiral BUZBY. That is something we would have to kind of look at. You know, that is a policy thing we would have to talk to the Secretary about for sure.

Mr. GARAMENDI. Well, you also talk to us about these things.

Admiral BUZBY. Certainly.

Mr. GARAMENDI. I think the chairman and I have made it clear where we are coming from on this. We think the MSP program is exceedingly important and that the subsidy level at $5.2 million was necessary.

Also, I note that the administration proposes to end the Public Law 480 program, which I think directly affects the MSP issue. I don't see the logic of that. I think that I don't see the logic of most of that budget, to be quite frank about it, except we are happy to have the additional money for the Coast Guard, as long as they spend that money appropriately on the things that we are talking about.

We talked earlier about the multimission training vessel and the $300 million. I want to be very, very clear where I am coming from.

I think this committee is supporting that position that we are not going to buy a foreign-built ship, we are not going to abandon the multimission.

So what does it cost to build those multimission ships?

Admiral BUZBY. A single vessel we estimate about $302 million.

If we were to build two or three, it drops down to about $280 million to $290 million.

Mr. GARAMENDI. So we would need somewhere around another $300 million over the course of the next couple of years to build two of those multimission ships.

What is the purpose of a multimission ship? What is the multimissions?

Admiral BUZBY. The reason that we are building the ship and calling it a National Security Multimission Vessel is its primary mission will be as a training vessel for the State maritime academy that it would be assigned to.

Its secondary mission would be, as we have seen this past summer, it would be used in a natural disaster relief sort of role, and as designed would have good capability to perform that role even better than our ships that were not purpose-built.

Mr. GARAMENDI. And these rusty hulks that are being discussed to be purchased, would they be able to do the multimission?

Admiral BUZBY. They would have some capabilities, clearly not as much as a ship that was purpose-built for that mission, no, sir.

Mr. GARAMENDI. So we have got to find a couple hundred million dollars to build the two ships and do it right the first time, right?

Correct?

Admiral BUZBY. That would be the best way to go forward, yes, sir.

Mr. GARAMENDI. I am sorry, that would be the——

Admiral BUZBY. The very best way to go forward.

Mr. GARAMENDI. The very best, most wonderful, beautiful way.

Admiral BUZBY. If we can all afford it, yes, sir.

Mr. GARAMENDI. I am learning the language of the administration.

We talked earlier about a big, beautiful, wonderful, border wall with steel that now has a 25-percent tariff on top of it, so the cost probably will end up some—but, in any case, that $18 billion. Admiral Zukunft, you said there were how many metric tons of cocaine did you seize on the low seas, the high seas?

Admiral ZUKUNFT. 233 this past year.

Mr. GARAMENDI. My information that I have received recently is that on the entire land border about 20 metric tons were seized.

That is about one-tenth of what you seized on the high seas.

Admiral ZUKUNFT. What is most vulnerable is where it moves in bulk, and that moves predominantly in the maritime domain way beyond our border.

Mr. GARAMENDI. So my message to the President is, from yesterday, he seemed to be really interested in stopping the importation of drugs, the illicit drugs coming into the Nation. And if he really wanted to get the best for the money, he would fund the Coast Guard. It is 10 to 1. You don't need to comment, but I put that out there.

Mr. President, we want to stop illegal drugs entering the United States. You want to spend your money in the very best, most efficient way possible. Take the $18 billion, spend some portion of it to rebuild the fences that we presently have. They can be big, beautiful fences or walls. But don't spend $18 billion that way. Spend it where it has the most effect, which is the U.S. Coast Guard.

And you need autonomous vehicles, drones and the like. You need better capability. You continue to work with the Latin American countries. All of this takes money.

And so if we want to get the very best, most efficient way to stop drugs coming into the country, it is the U.S. Coast Guard, not a beautiful wall paid for by the American taxpayers.

Enough of that. I think I am out of time once again, Mr. Chairman. Excuse me for being so blatant.

Mr. HUNTER. We actually did great compared to the other Members today.

Admiral Zukunft, I have one last question. The fiscal year 2017 approps, you had about $1.3 billion. This year, even though you are under a CR [continuing resolution] still, it will be $1.8 billion minus the months we have been in the CR, but it will be at $1.8 billion level, if everything goes the way it is supposed to go; 2019, you are almost at $1.9 billion. So you have 2 years at about $1.8 billion, $1.9 billion.

But 2020 it drops down to $1.4 billion. It doesn't go above $1.7 billion. It never reaches $2 billion at all going forward. Is that because we are just looking out basically 2 years at a time, or is that because of some other reason?

That is a precipitous dropoff. That is a $400 million drop.

Admiral ZUKUNFT. Right. Where you see the dropoff is in the polar icebreaker program. And this is at a point in time where we will have awarded final design, and we will be in a position to make an informed decision to do a block buy.

And so that is a concern when I see that precipitous of a drop, because that drop is mostly in the polar icebreaker building program. So that is a significant add-on that we are going to need going forward in order to look at——

Mr. HUNTER. So you are saying from 2020 forward, that does not include icebreaker funding?

Admiral ZUKUNFT. Well, where we see that significant drop, it really comes within the icebreaking account. That accounts for almost all of that reduction of nearly $500 million.

Mr. HUNTER. Gotcha.

OK. That is all I have.

Mr. Garamendi.

Mr. GARAMENDI. Two more very quick questions.

This is kind of like your mantra, Mr. Chairman, and you have gone through all of the reports that we have not received in the timeliness of those, and I appreciate you continuing to stay on that issue.

However, Admiral Buzby, there is one other report that we authorized or asked for in 2014. The previous administration didn't get it done, although I understand there was a draft that was held up in OMB. So where is the National Maritime Strategy?

Admiral BUZBY. Thank you for that question.

I am pleased to report that the National Maritime Strategy, in its revised state, has left my desk. It is now routing through my staff, prior to going up to the Secretary. And I believe it will be going very quickly.

It was a good piece of work. It needed just some touching up. I participated in its construction as a contributor earlier, prior to this current position.

So it is moving along. My goal is to deliver it to you, sir.

Mr. GARAMENDI. I think I have heard this before, with no offense to your good efforts. But please understand that we await that.

And your good efforts were repeated—actually you repeating the good efforts of your predecessor—and it got held up somewhere above your level.

Admiral BUZBY. Yes, sir.

Mr. GARAMENDI. We look forward to that.

Second question. It looks to me like the FMC isn't going to be able to have a quorum. This seems to be a problem that plagues the administration on appointments. I think Mr. Maffei's appointment ends in June and then you will be down to two people.

Mr. KHOURI. It is my understanding that even with two we still have a quorum. It is also my understanding that the Surface Transportation Board has operated with one and they continue to do business. I am not suggesting that that is recommended, optimal, or anything else, but our charter does not prevent us from continuing to operate.

I think each independent agency, you have to look at their individual charter to see what is required for a forum. But it is something we have thought about and looked into, Congressman.

Mr. GARAMENDI. I think by raising the question, I make my point. If one were to look at the White House, it appears as though there will be one person there also.

Thank you so very much. I yield back my time.

Mr. KHOURI. Thank you.

Mr. HUNTER. I thank the gentleman.

I thank all of you for being here and sharing with us and your testimony.

Admiral, again, Master Chief, whatever you guys say in the Coast Guard, fair wind, safe travels, good seas, all those things.

Thank you very much for your service to this Nation. It will be remembered. Thank you.

We are now adjourned.

[Whereupon, at 12:35 p.m., the subcommittee was adjourned.]

U.S. Department of Homeland Security
United States Coast Guard
Commandant
United States Coast Guard
2703 Martin Luther King Jr Ave SE
Washington, DC 20593-7000
Staff Symbol: CCG

TESTIMONY OF ADMIRAL PAUL F. ZUKUNFT COMMANDANT, U.S. COAST GUARD ON "THE COAST GUARD'S FISCAL YEAR 2019 BUDGET REQUEST," BEFORE THE HOUSE COAST GUARD AND MARITIME TRANSPORTATION SUBCOMMITTEE MARCH 14, 2018

Introduction

Mr. Chairman and distinguished members of the Committee, I appreciate the opportunity to testify today. Thank you for your enduring support of the United States Coast Guard, particularly the significant investments provided in the FY 2017 Consolidated Appropriations Act, recent Hurricane Supplemental, and ongoing deliberations to support our FY 2018 and FY 2019 President's Budget requests.

As the world's premier, multi-mission, maritime service, the Coast Guard offers a unique and enduring value to the Nation. The only branch of the U.S. Armed Forces within the Department of Homeland Security (DHS), a federal law enforcement agency, a regulatory body, a first responder, and a member of the U.S. Intelligence Community the Coast Guard is uniquely positioned to help secure the maritime border, combat transnational criminal organizations (TCO), and safeguard commerce on America's waterways.

The Coast Guard's combination of broad authorities and complementary capabilities squarely aligns with the President's national security and economic prosperity priorities; furthermore, it offers an agile toolset to address the Nation's most pressing challenges. Appropriately positioned in DHS, the Coast Guard is a military service and a branch of the Armed Forces of the United States at all times.[1] We are also an important part of the modem Joint Force[2] and currently have forces assigned to each of the five geographic Combatant Commanders, as well as Cyber Command.

---

[1] 14 U.S.C. § 1; 10 U.S.C. § 101.

[2] In addition to the Coast Guard's status as an Armed Force (10 U.S.C. § 101), see also Memorandum of Agreement Between the Department of Defense and the Department of Homeland Security on the Use of Coast Guard

As demonstrated in the 2017 record hurricane season, the Coast Guard is the Nation's "maritime first responder" and plays a leading role in executing the National Response Plan (NRP) for disaster situations. Our ability to rapidly surge in response to emerging threats or contingencies are critical to success across the spectrum of missions we prosecute.

We live in an increasingly volatile, uncertain, complex, and ambiguous world. Rapid technological advancement, increasing globalization, and intensifying threats from state and non-state actors alike challenge international norms and threaten global governance.

To ensure we meet the demands of today while preparing for tomorrow, the Coast Guard is guided by a five-year Strategic Intent and suite of regional and functional strategies that drive our Service's operations and investments.

These strategic efforts are informed by the National Security Strategy and applicable DHS strategies, and are coordinated to augment Department of Defense (DoD) priorities. Using these strategies as guideposts, leveraging the intelligence community, and employing a risk-based approach to focus our limited resources allows us to address maritime threats with the greatest precision and effect.

Strategic Effects

Fueled by the Service's unique authorities and capabilities, our Western Hemisphere Strategy continues to yield large-scale successes in our counter-drug mission. The Coast Guard's persistent offshore presence and associated interdiction efforts sever the supply lines of criminal networks where they are most vulnerable-at sea. Leveraging over 30 multilateral and bilateral agreements with a host of government organizations, the Coast Guard's long-term counter-TCO efforts promote stability and strengthen the rule of law throughout these regions. Working with interagency partners, the Coast Guard seized 223 metric tons of cocaine and detained and transferred 606 smugglers for criminal prosecution in FY 2017. Highlighting our record-breaking mission performance for drug interdiction was the STRATTON's offload of over 50,000 pounds of illicit narcotics, with an estimated street value of over $6.1 billion. This was a result of collaborative efforts between four U.S. Coast Guard cutters, DHS maritime patrol aircraft, and a U.S. Navy ship in over 25 separate interdictions. Beyond the important task of removing cocaine from the illicit system that gets it to U.S. streets, prosecuting smugglers facilitates deeper understanding of TCOs and ultimately helps our unified efforts to dismantle them.

Without question, National Security Cutters (NSC) have been a game-changer not only for our drug interdiction and counter-TCO operations in the southern maritime transit zone, but also in contributing to other national security priorities, such as

---

Capabilities and Resources in Support of the National Military Strategy, 02 May 2008, as amended 18 May 2010.

supporting DoD Combatant Commander requirements across the globe and projecting sovereign rights in the Arctic.

Looking forward, the Offshore Patrol Cutter (OPC) will provide the tools to more effectively enforce Federal laws, secure our maritime borders, disrupt TCOs, and respond to 21st century threats. Continued progress on this acquisition is absolutely vital to recapitalizing our aging fleet of Medium Endurance Cutters (MECs), some of which will be over 55 years old when the first OPC is delivered in 2021. In concert with the extended range and capability of the NSC and the enhanced coastal patrol capability of the Fast Response Cutter (FRC), OPCs will be the backbone of the Coast Guard's strategy to project and maintain offshore presence.

As one of the five Armed Forces, the Coast Guard deploys world-wide to execute our statutory Defense Operations mission in support of national security priorities. On any given day, II cutters, two maritime patrol aircraft, five helicopters, two specialized boarding teams, and an entire Port Security Unit are supporting DoD Combatant Commanders on all seven continents. In the Middle East, our squadron of six patrol boats continues to police the waters of the Northern Arabian Gulf in close cooperation with the U.S. Navy, promoting regional peace and stability. Likewise, as one of the principal Federal agencies performing detection and monitoring in the southern maritime transit zone, the Coast Guard provides more than 4,000 hours of maritime patrol aircraft support and 2,000 major cutter days to DoD's Southern Command each year.

In the high latitudes, the Arctic region is becoming increasingly accessible at a time when global interests in energy, clean water, and subsistence continue to intensity. The Coast Guard is committed to the safety, security, and environmental stewardship of the Arctic, and we will remain closely engaged with our partners, including Russia, via the Arctic Coast Guard Forum. By focusing on collaboration over conflict, we are promoting governance and building a shared approach to prevention and response challenges in the region.

Meanwhile, the 42-year old POLAR STAR recently completed another Operation DEEP FREEZE patrol in Antarctica. Just one major casualty away from leaving the Nation without any heavy icebreaking capability, POLAR STAR supported U.S. strategic interests and the National Science Foundation by breaking a navigable shipping lane to deliver fuel and critical supplies to the U.S. base at McMurdo Sound.

I appreciate your support for the $150 million appropriated in Shipbuilding and Conversion, Navy (SCN) funding in the FY 2017 Onmibus. This is a great step forward to secure our future in the Polar Regions and finally recapitalize the Nation's icebreaker fleet. This funding coupled with the $750 million in the FY 2019 President's Budget, would enable the Coast Guard to award a contract for detail design and construction and deliver the first new heavy polar icebreaker in 2023. These critical investments reflect our interests and standing as an Arctic Nation and affirm the Coast Guard's role in providing assured access to the Polar Regions.

At the same time the Service was conducting counter-drug missions in the Eastern Pacific and projecting sovereign rights in the Arctic, the Coast Guard also launched one of the largest responses in history during a historic 2017 hurricane season. Over a five week period, Hurricanes HARVEY, IRMA, MARIA, and NATE impacted over 2,540 miles of shoreline[3], and Coast Guard men and women in helicopters, boats, cutters, vehicles and on foot rescued over 11,300 people and over 1,500 pets.

During our 2017 hurricane response, the Coast Guard resolved over 1,269 aids to navigation discrepancies, handled 290 pollution cases, located and assessed more than 3,623 grounded vessels, with more than 1,585 removed to date. Within hours after each storm's passage, Coast Guard damage and recovery assessment teams were on-scene determining the status of ports and waterways, leveraging electronic aids to navigation when feasible to facilitate the rapid reopening of key ports and waterways, and assessing impacts to Coast Guard facilities and capabilities. This enabled a vital portion of the country's waterways to reopen, helping maintain our Maritime Transportation System (MTS) which contributes $4.6 trillion annually to our Gross Domestic Product.

The daily activities of Coast Guard men and women are heroic, as they support nearly every facet of the Nation's maritime interests, protect our homeland, and secure our economic prosperity. In addition to the hurricane responses, the Coast Guard prosecuted over 16,000 search-and-rescue cases and saved more than 4,200 lives; interdicted more than 2,500 undocumented migrants; completed over 9,100 Safety of Life at Sea safety exams on foreign vessels; and responded to over 12,200 reports of pollution incidents.

Beyond operations, we earned our fifth consecutive clean financial audit opinion the only Armed Service that can make such a claim. Further, our major acquisition programs and product lines are delivering new assets on schedule and on budget that have proven to meet our operational requirements. To better guide our modernization, we developed a Long Term Major Acquisitions Plan (LTMAP), a roadmap to field modem platforms to address 21'1 century threats. We have been working with the Administration to finalize the details of the LTMAP and are committed to delivering this report to Congress as soon as possible.

Our greatest strength is undoubtedly our people. Coast Guard operations require a resilient, capable workforce that draws upon the broad range of skills, talents, and experiences found in the American population. In FY 2019, the Coast Guard will maintain a proficient, diverse, and adaptable workforce that responds effectively to changing technology, an increasingly complex operating environment, and dynamic partnerships. Together, modem platforms and a strong, resilient workforce will maximize the Coast Guard's capacity to meet future challenges.

---

[3] Using CRS method of Shoreline Measurement: Texas: 367 mi, Louisiana: 397 mi, Florida: 1,350 mi, Puerto Rico: 311 mi, USVI: 117 mi.

Conclusion

History has proven that a responsive, capable, and agile Coast Guard is an indispensable instrument of national security. Funding 21st century Coast Guard platforms and people are especially prudent investments given today's challenging fiscal environment. I firmly believe no other investment will return more operational value on every dollar than the extraordinary men and women of the U.S. Coast Guard-which includes 48,000 Active Duty and Reserve members, 8,500 civilians, and over 27,000 volunteer members of the Coast Guard Auxiliary. As illustrated by our sustained response to an historic hurricane season, another record year removing illicit narcotics from the maritime approaches, and unique support to Combatant Commanders around the globe; our ability to rapidly surge resources to emerging threats continues yield unprecedented results for the Nation.

With the continued support of the Administration and Congress, the Coast Guard will continue to live up to our motto *Semper Paratus* -Always Ready. Thank you for all you do for the men and women of the Coast Guard.

FY 2019 Budget Request

The FY 2019 President's Budget funds the most critical Coast Guard operations and continues our Service's highest priority recapitalization efforts. The budget efficiently allocates resources to optimize performance across all Coast Guard mission programs and activities. The Coast Guard must continue to meet today's operational requirements while investing in future capability to best serve the Nation.

The Coast Guard's FY 2019 budget request is focused on three main priorities:

1) Invest in the 21st Century Coast Guard
2) Sustain Mission Excellence
3) Maximize Value to Nation

Invest in the 21st Century Coast Guard

Coast Guard mission demands continue to grow and evolve. The complexities and challenges facing our Nation require well-trained Coast Guard men and women with capable platforms providing the persistent presence necessary to conduct operations. Given the age and condition of the Coast Guard's legacy assets, future mission success relies on continued recapitalization of Coast Guard boats, cutters, aircraft, systems, and infrastructure.

The FY 2019 President's Budget funds construction of the second OPC and long lead time material (LLTM) for the third. OPCs will be the cornerstone of our Service's future surface fleet and comprise 70 percent of the Coast Guard's offshore presence. To meet growing demands in the Polar Regions, the budget provides funding to award a contract for detail design and construction of polar icebreakers and maintains the critical path to deliver the first new heavy polar icebreaker in 2023. Just as critical, the budget funds the service life extension project for the POLAR STAR, the Nation's only heavy icebreaker in service. The budget also continues other vessel acquisition programs, providing funds for four FRCs, totaling 52 hulls of the complete program of record of 58 hulls. In addition to surface recapitalization efforts, the FY 2019 President's Budget continues sustainment and conversion work on in-service fixed and rotary wing aircraft, including missionization of the C-27J aircraft received from the Air Force, and investment in Small Unmanned Aircraft Systems (sUAS), which have proven to be highly successful in achieving strategic objectives in the Western Hemisphere.

Sustain Mission Excellence

The FY 2019 President's Budget ensures the Coast Guard can conduct today's highest priority operations in support of national objectives. Most importantly, it sustains the Coast Guard's workforce and supports proficiency, maximizing operational safety and effectiveness.

The FY 2019 President's Budget provides critical operations and maintenance funding for more capable, modernized assets delivered via the Coast Guard's acquisition programs, including one NSC, six FRCs, and two HC-130J aircraft. In all, the FY 2019 President's Budget increases the workforce by 93 FTE to support Coast Guard activities across all mission programs. The budget also maintains parity with DOD for military pay, allowances, and health care, and for civilian benefits and retirement contributions.

Maximize Value to Nation

In best serving the Nation, the Coast Guard must continue to meet evolving mission requirements stemming from national priorities and remain a trusted steward of public resources.

The 2019 President's Budget sustains the most critical frontline operations by efficiently allocating resources across all mission programs. Coast Guard Operational Commanders will maintain search and rescue coverage, protect critical infrastructure, counter illicit threats from entering the United States, facilitate safe navigation and commerce within the vital MTS, safeguard the maritime environment, support foreign policy objectives, and conduct defense operations in support of DOD Combatant Commander requirements.

FY 2019 Budget Highlights

Procurement, Construction, & Improvements (PC&I)

Surface Assets

The budget provides $1,544 million for the following surface asset recapitalization and sustainment initiatives:

- NSC-Provides funding for post-delivery activities for the seventh through ninth NSCs, and test and evaluation activities. The acquisition of the NSC is vital to performing DHS missions in the far offshore regions, including the harsh operating environments of the Pacific Ocean, Bering Sea, and Arctic. The NSC also provides a robust command and control platform for homeland security and contingency operations.
- OPC-Provides funding for construction of the second and LLTM of the third OPC, which are scheduled for delivery in 2022 and 2023, respectively. The OPC will replace the Medium Endurance Cutter classes, now well beyond their service lives, which conduct missions on the high seas and coastal approaches.
- FRC-Funds procurement of four FRCs, totaling 52 of the complete program of record of 58. These assets replace the less capable 110-foot patrol boats, enhancing the Coast Guard's coastal capability to conduct Search and Rescue operations, enforce border security, interdict drugs, uphold immigration laws, prevent terrorism, and enhance resiliency to disasters.
- Polar Icebreaker-Provides funding to award a contract for detail design and construction of polar icebreakers in 2019, delivering the first new heavy polar icebreaker in 2023. New heavy polar icebreakers will provide the Nation with assured surface access to the Polar Regions for decades to come.
- Polar Sustainment Supports program management activities, survey and design efforts, and LLTM purchases for a multi-year Service Life Extension Project for POLAR STAR to provide surface presence in the Polar Regions.
- Waterways Commerce Cutter (WCC)-Provides funding for acquisition planning activities to continue evaluation for the replacement of the multi-mission platform integral to the protection of maritime commerce on the inland rivers.
- Cutter Boats-Continues funding for the production of multi-mission cutter boats that will be fielded on the Coast Guard's major cutter fleet, including the NSC.
- In-Service Vessel Sustainment Continues funding for sustainment projects on 140-foot Ice Breaking Tugs, 225-foot Seagoing Buoy Tenders, 270-foot Medium Endurance Cutters, and 47-foot Motor Lifeboats.

- Survey and Design-Continues funding for multi-year engineering and design work for multiple cutter classes in support of future sustainment projects. Funds are included to plan a Mid-Life Maintenance Availability (MMA) on the CGC HEALY.

Air Assets

The budget provides $148 million for the following air asset recapitalization or enhancement initiatives:

- HC-27Funds continued missionization activities of the C-27J, including funding for spare parts, logistics, training, and mission system development.
- HC-144 Funds continued Minotaur mission system retrofits and provides high-definition electro-optical infrared cameras to meet DHS Joint Operational Requirements.
- HH-65Continues modernization and sustainment of the Coast Guard's fleet ofH-65 short range recovery helicopters, converting them to multi-mission MH-65E variants.
- The modernization effort includes reliability and sustainability improvements, where obsolete components are replaced with modernized sub-systems, including an integrated cockpit and sensor suite. Initial funding is also included to extend aircraft service life for an additional 10,000 hours.
- MH-60Includes initial funding to extend aircraft service life for an additional 10,000 hours to better align with DOD's H-60 replacement timeline.
- sUAS-Continues program funding to deploy sUAS onboard the NSC allowing increased interdiction through greater Intelligence, Surveillance, and Reconnaissance (ISR).

Shore Units and Aids to Navigation (ATON)

The budget provides $135 million to recapitalize shore infrastructure that supports Coast Guard assets and personnel, as well as construction and improvements to ensure public safety on waterways:

- Specific Projects - Funds continued renovation and modernization of Chase Hall Barracks at the Coast Guard Academy, site work and construction of a new residential complex in Wailupe, Hawaii, and other minor repairs and improvements.
- Major Acquisition Systems Infrastructure Funds modification and construction of facilities to support newly arriving assets. Includes upgrades and construction for

NSC, OPC, and FRC homeports and also supports upgrades for Air Station Barbers Point, Hawaii to transition to C-130Js.

Other (Asset Recapitalization)

The budget provides $60 million for other initiatives funded under the Procurement, Construction, and Improvements account, including the following equipment and services:

- Command, Control, Communications, Computers, Intelligence, Surveillance, and Reconnaissance (C4ISR)Provides design, development, upgrades, and assistance on C4ISR hardware and software for new and in-service assets.
- Program Oversight and Management-Funds activities associated with the transition of the Coast Guard's assets from acquisition to operations, including delivery, provision of logistics, training, and other services necessary to ensure seamless integration into the operational fleet.
- CO-Logistics Information Management System-Continues development and deployment of this system to Coast Guard operational assets.
- Other Equipment and Systems-Funds end-use items costing more than $250,000 used to support Coast Guard missions, including equipment to support operation and maintenance of vessels, aircraft, and infrastructure.

Operations and Support (0&8)

Operation and Maintenance of New Assets

The budget provides $63 million and 189 FTE to operate and maintain shore facilities and sustain new cutters, boats, aircraft, and associated C4ISR subsystems delivered through acquisition efforts:

- Shore Facilities - Funds operation and maintenance of shore facility projects scheduled for completion prior to FY 2019.
- FRC-Funds operation and maintenance of FRCs #31-36, as well as personnel for FRC crews #34-38 and shore-side support for FRC homeports in Cape May, NJ; Galveston, TX; Honolulu, HI; San Juan, PR; and San Pedro, CA.
- NSC-Funds operations and maintenance of NSC #8, as well as personnel for sensitive compartmented information facility (SCIF) crews and analytical support, and shore-side support for a third NSC homeport in Honolulu, HI.
- C-27J Aircraft-Funds fixed-wing aircraft maintenance support personnel.
- HC-130J Aircraft-Funds operations, maintenance, air crews, and pilots for HC-130J airframes #10-11.

Pay & Allowances

The budget provides $101 million to maintain parity with DOD for military pay, allowances, and health care, and for civilian benefits and retirement contributions. As a branch of the Armed Forces of the United States, the Coast Guard is subject to the provisions of the National Defense Authorization Act, which include pay and personnel benefits for the military workforce.

Asset Decommissionings

The budget saves $16 million and 56 FTE associated with the planned decommissioning of Coast Guard aircraft. As the Coast Guard recapitalizes its cutter and aircraft fleets and brings new assets into service, the older assets that are being replaced will be decommissioned:

- HC-130H Aircraft - The budget decommissions four HC-130H aircraft. These assets are being replaced with modernized and more capable HC-130J aircraft.

Operational Adjustments

In FY 2019, the Coast Guard will make sound, risk-based operational decisions while investing in critical recapitalization and new workforce initiatives:

- Aircraft Federal Aviation Administration (FAA) Compliance - The budget provides $20 million to replace obsolete aircraft equipment and systems necessary to comply with FAA 2020 airspace requirements.
- Obsolete Equipment Replacement - The budget provides $11 million and 1 FTE to replace obsolete equipment on cutters and aircraft, including: military satellite communications and cutter underway connectivity equipment.
- Mission Essential Systems - The budget provides $5 million to sustain critical network infrastructure necessary to comply with DOD information network and cybersecurity requirements.
- Acquisition Personnel and Management - The budget provides $7 million and 46 FTE to increase the Coast Guard's professional acquisition workforce to support new and existing recapitalization projects and programs.
- Elimination of the Crew Rotation Concept (CRC) Pilot Program The budget saves $32 million and 231 FTE associated with the elimination of CRC operations on NSC hulls 1-3 homeported in Alameda, California. Elimination of the CRC program standardizes operations across the NSC fleet and avoids costly and inefficient CRC implementation on other NSCs.

Question#:	I
Topic:	Request for Information
Hearing:	Review of Fiscal Year 2019 Budget Request for the Coast Guard and Maritime Transportation Programs
Primary:	The Honorable Duncan D. Hunter
Committee:	TRANSPORTATION (HOUSE)

Question: On March 1, 2017, the U.S. Coast Guard (USCG) issued a Request for Information (RFI) for commercially available Boat Crew Communications System (BCCS). THE RFI calls for an untethered, hands-free wireless intercom capability, hardened for the maritime environment. It is my understanding that USCG conducted demonstrations June 12-14, 2017. Please provide the Subcommittee with an overview of the results of that demonstration; a timeline for a formal Request for Proposal, and the anticipated evaluation and acquisition timeline.

Response: In response to the RFI, the Coast Guard received capability statements from seventeen companies. Twelve companies elected to present their solutions to the government. Many were able to demonstrate wireless systems that appear to meet the Coast Guard's need. Based on industry input, the Coast Guard was able to refine the requirements and plan for competitive procurement. The Coast Guard intends to release a Request for Proposal during the second half of 2018.

Question: The Coast Guard has issued other RFI's within the last two years for aviation platform (fixed wing and rotary wing) wireless intercom capability. Has USCG considered an effort to standardize hands-free wireless intercom capability across all or most of the USCG's maritime and aviation platforms?

Response: Yes. Standardization is considered as part of the normal engineering design process. The Coast Guard evaluates its requirements for crew communications capability as part of obsolescence management for each platform type, including aviation and afloat.

Question: If so, how does the planned BCCS program align with that effort?

Response: BCCS requirements will be compared to other enterprise requirements and solutions to determine potential for standardization and efficiency. Identifying existing enterprise solutions is a routine step in meeting the needs of emerging Coast Guard capabilities.

Question#:	2
Topic:	Suspect or Counterfiet Items
Hearing:	Review of Fiscal Year 2019 Budget Request for the Coast Guard and Maritime Transportation Programs
Primary:	The Honorable Rick Larsen
Committee:	TRANSPORTATION (HOUSE)

Question: During a recent investigation of a ferry propulsion failure, the Coast Guard discovered evidence of the use of falsely identified components. As a result, last April you issued a safety alert "strongly recommend[ing] that all vessel owners, operators, and technical personnel maintain awareness Suspect *I* Counterfeit Items (S/Cis) and to implement policies and procedures aimed at identifying them."

I recently met with a company in my district that offers high-resolution imaging technology used to detect counterfeit components in the military supply chain. Has the Coast Guard evaluated the use of such technologies in the maritime realm?

Response: The Coast Guard has not examined high resolution imaging technology for use in the maritime supply chain. With the exception of Coast Guard approved equipment, we do not directly audit the records of marine component manufacturers and suppliers. Vessel owners are responsible to ensure contracted suppliers of components meet original equipment manufacturer (OEM) or builders' drawing requirements.

Coast Guard approved lifesaving and firefighting equipment or systems must comply with strict procedures for their manufacture, modification, or repair. During vessel construction, or when a vessel undergoes significant repairs or modifications, the Coast Guard reviews and approves the plans for engineering systems and structural components to ensure they comply with an "as built" condition. The Coast Guard also verifies that engine or propulsion system components comply with design drawings and/or the OEM specifications. The Coast Guard encourages owners to use suppliers as authorized by the OEM.

Question: How would the President's fiscal year 2019 budget impact the Coast Guard's ability to proactively remove S/Cls from the maritime supply chain?

Response: The Coast Guard does not perform this function. Vessel owners are responsible to ensure the components they purchase for new construction or contracted repairs are proper OEM components.

Question#:	3
Topic:	Future VAS Use
Hearing:	Review of Fiscal Year 2019 Budget Request for the Coast Guard and Maritime Transportation Programs
Primary:	The Honorable Rick Larsen
Committee:	TRANSPORTATION (HOUSE)

Question: Last year in your State of the Coast Guard address, you highlighted the Coast Guard's lack of "eyes in the sky" as a major shortfall. With the emergence of new technology and the administration's interest in Unmanned Aircraft Systems (UAS) innovation, it is likely the Coast Guard's use of UAS will expand.

How will the President's proposed $17.2 million (or 47.4 percent) cut for research and development impact the Coast Guard's current or future use of UAS?

Response: The funds requested for R&D in the FY 2019 President's Budget do not constitute a cut in R&D funding. The funds appropriated to the Coast Guard in FY 2017 for the Long Range/Ultra Long Range UAS project were above the Coast Guard's base R&D funding level and will remain available through FY 2019. Therefore, this work will continue as planned and the FY 2019 U.S. Coast Guard R&D request does not represent a change in the Coast Guard's UAS program.

Statement of Mark Buzby, Administrator, Maritime Administration, U.S. Department of Transportation, before the Committee on Transportation and Infrastructure, Subcommittee on Coast Guard and Maritime Transportation, U.S. House of Representatives, Hearing on "the Maritime Administration's Fiscal Year 2019 Budget Request," March 14, 2018

Good morning, Chairman Hunter, Ranking Member Garamendi and members of the Subcommittee. I appreciate the opportunity to discuss the President's Fiscal Year (FY) 2019 budget priorities for the Maritime Administration (MARAD). MARAD's statutory mission to foster, promote, and develop the merchant maritime industry of the United States. This budget request furthers that mission by investing in U.S. mariner training, supporting programs that help U.S.-flag commercial vessels compete globally, and maintaining sealift readiness to meet national security requirements.

FY 2019 Budget Request

The United States is a maritime Nation, with our country's economy highly dependent on viable coastal and ocean resources, and inland waterways. In 2014, U.S. maritime freight activity helped to support $4.6 trillion of economic activity among the many non-maritime industries that depend on it for access to world markets and within the maritime industry itself, equivalent to 26 percent of our GDP.[4]

MARAD's programs seek to support both the maritime industry's commercial health and national security objectives.

[4] Martin Associates, "The 2014 National Economic Impact of the U.S. Coastal Port System," prepared for the American Association of Port Authorities, March 2015, p.G.

The U.S.-flag fleet operating in international trade currently carries less than 2 percent of our annual foreign trade by tonnage[5] The U.S. commercial presence in the international maritime domain has been at historic lows over the past several years with only 81 ships remaining in those trades. As our national economy relies on foreign trade, we need to increase our capability to participate as a nation in maritime commerce to ensure we remain economically competitive. Without this capability, the U.S. economy would become reliant on foreign-flag shipping services. During Operation Desert Shield in 1991, DOD found it necessary to employ 177 foreign vessels to meet sealift needs in addition to approximately 170 U.S.-flagged vessels. However, 13 of the foreign vessels carrying essential supplies hesitated or refused to enter the area of operations, resulting in a loss of 34 transit days for ships carrying cargo for U.S. troops.[6]

Additionally, Commander, United States Transportation Command, recently testified regarding the critical importance of the availability of U.S.-flag vessels using this example stating, "foreign-flagged vessels declined to enter the area of operations, while U.S.-flagged vessels provided steadfast support, delivering the bulk of the Joint Force to the Middle East."[7] As the fleet dwindles, so does the base of U.S. Merchant Mariners and the shipbuilding and repair industry, which are all essential components of national security.

To ensure a strong domestic maritime industry and U.S. Merchant Marine into the future, the President's FY 2019 Budget Request for MARAD of$696.4 million is focused on increasing the competitiveness of the U.S.-flag fleet, and investing in the education and training of the next generation of merchant mariners. MARAD remains committed to marine transportation policies that improve security, address our Nation's critical maritime infrastructure gaps, and leverage technology to meet the needs and challenges of the marine transportation industry. This request supports a variety of MARAD initiatives involving ships and shipping, vessel operations, national security and strategic mobility, ship disposal, and maritime education. A summary of the FY 2019 request is provided below.

National Security

By law, and pursuant to Presidential National Security Directive 28, DOD must rely on U.S.-flag ships crewed by volunteer, civilian U.S. mariners, to provide the sealift

---

[5] MARAD calculation of tonnage based on business confidential data from CBP and Census combined with commercially available registry data for the 2016 calendar year.

[6] So Many, So Much, So Far, So Fast: United States Transportation Command and Strategic Deployment for Operation Desert Shield/Desert Storm/ James K. Matthews, Cora J. Holt, p. 136.

[7] Statement of General Darren W. McDew, United States Air Force, Commander, United States Transportation Command before the Senate Armed Services Committee on the State of the Command, February 13, 2018. p. II.

capacity to support military deployments and respond to national emergencies. MARAD is charged with ensuring that U.S.-flag ships and merchant mariners are available to meet DOD sealift requirements. The U.S.-flag fleet of privately owned, commercially operated vessels, along with government-owned vessels, provide a critical public-private sealift surge and sustainment capacity to move equipment and materials for the Armed Forces and Federal agencies as needed, during times of conflict, humanitarian crises, and natural disasters such as those we witnessed last summer in the wake of Hurricanes Harvey, Irma, and Maria.

Maritime Security Program (MSP)

For FY 2019, $214 million is requested for the MSP, providing for $3.6 million for each of the 60 ships enrolled in the program. The Maritime Security Act of 1996 established the MSP, which ensures access to U.S.-flag ships in ocean-borne foreign commerce and the necessary intermodal1ogistics networks to move military equipment and supplies during armed conflict or national emergency. This program also provides critical employment for up to 2,400 highly qualified U.S. merchant mariners and approximately 5,000 shore side maritime professionals each year. The MSP provides direct annual stipends for 60 active, commercially viable, militarily useful, privately owned U.S.-flag vessels and crews operating in the international trade. Participating operators commit their ships, crews, and global network of intermodal facilities and commercial transportation resources upon request by the Secretary of Defense during times of war or national emergency.

During FY 2018, MARAD restored the MSP fleet to the full, authorized level of 60 ships. Being at full capacity bolsters the ability of the U.S. merchant marine to meet DOD deployment and sustainment requirements. MARAD brought two new roll-on roll-off (RO/RO) vessels into the MSP fleet, which added more than 320,000 square feet in militarily useful cargo capacity, significantly improving DOD's ability to move and sustain the heaviest U.S. armored units.

Entry of the new ships raised the overall fleet's militarily useful capacity to the highest level in the program's history. The MSP has supported every U.S. conflict since its inception in 1996, including Operations Enduring Freedom and Iraqi Freedom. These vessels stand ready to play a vital role in support of U.S. military operations worldwide.

National Defense Reserve Fleet (NDRF) and Ready Reserve Force (RRF)

MARAD maintains a fleet of government-owned merchant ships in the NDRF. This includes 46 RRF vessels that are maintained and ready for operation within five or ten days for transport of military cargo to critical areas of operation. Our Nation has called upon RRF and NDRF vessels, which include training ships on loan to the six state maritime academies (SMAs) and the U.S. Merchant Marine Academy (USMMA), to respond to several disasters. RRF and NDRF ships were activated to provide support to other government agencies for recent relief efforts following Hurricanes Harvey, Irma, and Maria. During these deployments, these vessels supplied first responders with housing, meals, logistical support, and relief supplies, including critical replacement of Federal Aviation Administration air navigation equipment that was delivered by one of the activated vessels to the Virgin Islands. Additionally, these vessels were previously activated for disaster relief following Hurricanes Katrina, Rita, and Sandy and the 2010 earthquake relief effort in Haiti.

In addition to hurricane response, this past year MARAD supported DOD with the largest round of ammunition cargo sealifts in decades, as well as a 57-day mission for an Army unit movement to Kuwait where five vessels were utilized to accomplish six distinct sealift missions. Along with cargo sealift activities, two MARAD vessels were utilized by the DOD's Missile Defense Agency for tracking multiple missile test launches.

Funding provided from DOD will allow MARAD to continue to provide ready surge sealift support in FY 2019 and special mission vessels from the RRF fleet, while maintaining MARAD's NDRF fleet mooring sites. This request includes an additional increase in funding that will support readiness related maintenance and repair initiatives for service life extension work supporting the special mission vessels. MARAD is working with the U.S. Transportation Command (USTRANSCOM) and the U.S. Navy to address the urgent need for recapitalization of the RRF to ensure the readiness of these 46 ships, the average age of which is 43 years.

Maritime Training

MARAD provides funding and oversight for mariner training programs to produce highly skilled U.S. Coast Guard (USCG) credentialed officers for the U.S. Merchant Marine. It takes many years of training to develop the necessary mariner competencies for deck and engineering officer positions on large vessels in international trade. Therefore, maintaining an adequate pool of U.S. merchant mariners is vital to both the peacetime commercial success of the U.S.-flag fleet and to maintain the capacity needed to man Government-owned surge sealift forces to deploy and sustain US forces overseas in times of national emergency. The USMMA and SMAs graduate most of the USCG-credentialed officers who hold an unlimited tonnage or horsepower endorsement available to crew these U.S.-flag ocean-going ships. These graduates support our Nation with a cadre of well-educated and trained merchant mariners capable of serving in support of military operations, national emergencies, and humanitarian missions.

United States Merchant Marine Academy

The President's FY 2019 Budget Request includes $74.6 million for USMMA. Of this, $70.6 million will support Academy operations, and $4 million will fund priority capital repairs and improvements to the Academy's buildings and infrastructure. This request will enable the Academy to effectively achieve its core responsibility to educate and train the next generation of outstanding leaders as shipboard officers at sea and commissioned officers in our active and reserve armed forces. The USMMA is an accredited institution of higher education operating under the DOT and managed by MARAD. The USMMA offers a four-year maritime-focused program, centered on rigorous academic and practical technical training that leads to a Bachelor of Science degree, a USCG merchant mariner credential (MCC) with an unlimited tonnage or horsepower officer endorsement, and a commission as an officer in the reserve or active Armed Forces. Distinctly, USMMA graduates incur an obligation to serve five years as a merchant marine officer aboard U.S. documented vessels or on active duty with the U.S. Armed Forces or uniformed services. If serving in the reserves, they must remain as a commissioned officer for eight years. In June 2019, 196 midshipmen are expected to graduate from the Academy.

DOT, MARAD, and the USMMA take sexual assault and sexual harassment at the Academy and at sea very seriously, and have worked collaboratively with industry stakeholders to facilitate a cultural change by disseminating a strong message against sexual assault and sexual harassment in the maritime sector. The Academy is implementing provisions included in both the Fiscal Year 2017 and Fiscal Year 2018 National Defense Authorization Act aimed at improving the Academy's sexual assault and sexual harassment prevention and response efforts. Actions include enhancing prevention training, increasing campus security, initiating a Midshipman- driven on-campus culture change program, and expanding the Sexual Assault Prevention and Response Office. The FY 2019 funding will also provide for satellite communication devices that will be issued to midshipmen during Sea Year training and upgrading the 24/7 sexual assault hotline.

State Maritime Academies

In addition to providing oversight of the USMMA, MARAD provides funding assistance to six SMAs,[8] which collectively graduate more than three-fourths of the entry-level merchant marine officers annually. The President's FY 2019 Budget Request includes $24.4 million for the SMA program. This request includes $22 million to fund maintenance and repair costs for Federally owned school ships on loan from MARAD to

[8] The six SMAs are: California Maritime Academy in Vallejo, CA; Great Lakes Maritime Academy in Traverse City, MI; Texas A&M Maritime Academy in Galveston, TX; Maine Maritime Academy in Castine, ME; Massachusetts Maritime Academy in Buzzards Bay, MA; and State University of New York (SUNY) Maritime College in the Bronx, NY.

the SMAs. Ensuring the continued availability of SMA training vessels is a critical need and high MARAD priority. Training ship maintenance work is increasingly critical and costly as the ships age, and approach or exceed their designed service life. MARAD will use the funds to address priority maintenance needs across all the training vessels.

Additionally, the request provides $2.4 million to fund the Student Incentive Program (SIP), which provides financial assistance to 75 new cadets each year (across all SMAs) that participate in the program for a period of four years. The SIP program provides cadets with funds for uniforms, tuition, books, and subsistence. Upon graduation, SIP students must maintain an unlimited USCG credential for six years, fulfill a three-year service obligation in the maritime industry, and serve in a reserve unit of an Armed Forces or uniformed service for eight years. The SMAs regard the SIP program and support for their training ships as among the most important recruiting tools to encourage potential cadets to pursue careers as merchant marine officers. Approximately 991 Cadets are expected to graduate from the SMAs in May 2019.

School Ship Replacement

The President's FY 2019 Budget Request includes $300 million to support the procurement and conversion of two used ships to replace aging and outdated training ships on loan from MARAD to the SMAs. Specifically, this funding will be prioritized for replacing the Training Ship (TS) EMPIRE STATE, currently assigned to the State University of New York (SUNY) Maritime College, and TS KENNEDY, currently assigned to the Massachusetts Maritime Academy.

These two vessels are each more than 50 years old, and are now serving beyond their designed service lives. Both ships combined provide nearly 68 percent of the state maritime training capacity.

Investment in school ship replacement is essential to foster the growth of the Nation's maritime transportation workforce needs, while addressing emerging workforce challenges facing a critical shortage of U.S. merchant mariners needed to crew the commercial and government- owned sealift ships to meet national security needs. The Department is working to update its 2019 Budget materials to reflect the $300 million school ship request and will submit that revised information to Congress soon.

Other Maritime Transporation Programs

Ship Disposal Program

The President's FY 2019 Budget requests $30 million for the ship disposal program, of which $25 million is to continue, under Phase II, the radiological decontamination, dismantlement, and disposal of the defueled nuclear power plant on board the Nuclear Ship SAVANNAH (NSS). Funding also includes $3 million to maintain the NSS in protective storage, which supports and manages the Nuclear Regulatory Commission (NRC) required licensed activities on board the ship, including radiological protection, vessel maintenance, lay berthing, and custodial care. The ship disposal funding request also includes $2 million for salaries and administrative program costs.

As a Federal licensee, MARAD is responsible for continuing the required protective storage activities for the NSS until decommissioning and license termination are complete. The program received funding for decommissioning in FY 2017 to initiate and complete Phase I of a three- phased decommissioning project. Phase I is comprised of administrative and industrial activities that complete the prerequisites for commencement of the heavy engineering and industrial activities in Phase II. This FY 2019 request allows for a seamless decommissioning project transition, without shutdown interruption, into Phase II, which will be largely confined to activities involved with removing the reactor and major components. The NRC allows a maximum of 60 years from permanent nuclear power plant shutdown to license termination. All decommissioning and license termination activities must be completed by December 2031.

MARAD is also the ship disposal agent for Federal government-owned merchant-type vessels of 1,500 gross tons or greater and has custody of a fleet of non-retention ships. When ships are determined to be no longer of sufficient value to merit the cost of further preservation, MARAD arranges for their responsible disposal, with priority emphasis on the disposal of non-retention vessels in the worst condition. Currently, MARAD has 13 obsolete NDRF vessels not yet under contract for disposal, which is a historic low. In July 2017, MARAD completed the final removal of the remaining two non-retention vessels from the Suisun Bay Reserve Fleet, in compliance with an April2010 U.S. District Court consent decree requiring the removal of all 57 vessels from the fleet site by September 2017.

Conclusion

These programs represent MARAD priorities supported by the President's FY 2019 Budget request. We will continue to keep this Subcommittee apprised of the progress of our program activities and initiatives in these areas in the coming year.

Mr. Chairman, thank you for the opportunity to present and discuss the President's FY 2019 Budget Request for MARAD. I appreciate the Subcommittee's continuing support for maritime programs and I look forward to working with you on advancing maritime transportation in the United States. I will be happy to respond to any questions you and the members of the Subcommittee may have.

"Review of Fiscal Year 2019 Budget Request for Coast Guard and Maritime Transportation Programs" Subcommittee on Coast Guard and Maritime Transportation Hearing, Wednesday, March 14, 2018, Washington, D.C.

Questions for the Record

Question: The President's fiscal year 2019 budget eliminates the Small Shipyards Grants which supports facilities' modernization and repair efforts, and helps them compete globally. In the district I represent, Dakota Creek Industries in Anacortes received over $280,000 through the program.

1. How is the administration able to find $300 million in new funding for the procurement and retrofitting of used cargo ships (under the budget addendum), but cannot find simply three percent of that amount to continue the Small Shipyards grant program?

Response: MARAD's mission is to support the U.S. maritime infrastructure which includes maintaining the health of the U.S. Merchant Marine as well as the ship building industry. MARAD has invested nearly $187 million since FY 2008 to support shipyards such as Dakota Creek Industries through the Small Shipyards Grant program. An additional $20 million was recently appropriated for the program for FY 2018. These new grant awards will be announced later this summer.

The FY 2019 budget request reflects the current budget environment which required the Administration to make some difficult programmatic funding decisions. One of the major unaddressed challenges to the U.S. maritime infrastructure is the need to replace the aging school ship fleet that provides training to our future merchant mariners at the State Maritime Academies. The largest ship, SUNY Maritime's *TS EMPIRE STATE* will age out in 2019. Massachusetts Maritime's *TS KENNEDY* will age out in 2024. Both of these vessels combined provide nearly 69 percent of the State Maritime mariner training capacity. The Administration has been working hard to find solutions to address this challenge, knowing that it potentially requires a large capital investment from the federal government. The recent passage of the Bipartisan Budget Act of 2018 (Public Law 115-123) provided an opportunity for the Administration to request the funding needed to begin addressing this significant unmet need.

2. How does the Maritime Administration plan to support the efforts of small shipyards in the absence of this vital federal funding source?

Response: In FY 2018, MARAD will continue supporting the efforts of small shipyards with $20 million for the Small Shipyards Grant program. The solicitation for applications has been posted on the Federal Register, with applications from the shipyards due by May 22nd. MARAD will be awarding the grants later this summer.

In FY 2019, MARAD will continue to support the efforts of small shipyards by focusing on administering the existing grant portfolio of Small Shipyard Grant award recipients and providing the technical assistance they need to make their projects successful. Additionally, MARAD will continue to work with the entire U.S. maritime industry, including small shipyard owners, in identifying cost-effective solutions that support maintaining the maritime industry's commercial health and competitiveness around the world, and to advance marine transportation policies that improve security, address our Nation's critical maritime infrastructure gaps, and leverage technology to meet the needs and challenges of the marine transportation industry.

Statement of Michael A. Khouri, Acting Chairman Federal Maritime Commission, before the Committee on Transportation and Infrastructure, Subcommittee on Coast Guard and Maritime Transportation, United States House of Representatives, March 14, 2018

Chairman Hunter, Ranking Member Garamendi, and Members of the Subcommittee, thank you for providing me with this opportunity to appear before you today and discuss the mission of the Federal Maritime Commission. I appreciate the opportunity to share with you how the Commission works to safeguard competition in ocean transportation for the benefit of the American consumer. I want to acknowledge my fellow Commissioners here today Commissioners Dye and Maffei.

The Federal Maritime Commission

The FMC is an independent agency with specialized expertise that administers an antitrust regulatory regime tailored to address the unique characteristics and issues affecting the international ocean liner trade. Since 1916, Congress has recognized that the ocean liner industry, which transports a large percentage of the exports and imports so essential to this Nation's commerce, requires particular consideration because of the sector's critical role in our economy, its international dimension, and the competing and potentially conflicting regulatory regimes and interests of our partners.

The Federal Maritime Commission is a competition enforcement agency with a mission to ensure a competitive and reliable international ocean transportation supply system that supports the U.S. economy and protects the public from unfair and deceptive practices. The Commission facilitates an open and free market for ocean shipping services and protects against anticompetitive behaviors. It is vital to the interests of all parties involved in the movement of ocean commerce that the Commission provide confidence that America has a competitive, fair, and efficient ocean shipping marketplace.

Our Annual Report is scheduled to be submitted by April I, 2018, and will provide you with a comprehensive summary of the Commission's activities and industry developments in Fiscal Year 2017 (FY 2017). I will briefly review our activities of the past year, address what we foresee as potential developments and trends in the coming months, and discuss our Fiscal Year 2019 (FY 2019) budget request.

Changes from 2016 to 2017 and Industry Oversight

The container shipping industry plays an integral role in America's international trade and commerce. In 2017, approximately 34 million TEUs[9] moved through our ports, a 4 percent increase from 2016. U.S. imports surged during the year and accounted for most of this increase. The U.S. imported over 22 million TEUs last year valued at $754 billion. This was an increase of over 6 percent by volume from 2016. Meanwhile, the U.S. exported 12 million TEUs in 2017 with a value of $266 billion, a 1 percent increase over 2016 by volume.

In 2016, there were significant changes to the ocean transportation services marketplace, marked by merger and acquisition activity among shipping lines and the bankruptcy of a top ten ocean carrier. As a result of these events, the number of major multi-trade lane shipping lines operating in the international trades has dropped from 21 in 2011 to 12 global carriers following the merger of the three Japanese carriers into Ocean Network Express (ONE) and COSCO's acquisition of OOCL. The table below lists the ocean carriers that serve the major east-west trade lanes. On a broader scale, thirty-six carriers serve the U.S. trades.[10]

Notwithstanding the reduction in the number of major shipping lines serving the international trades, the container industry remains very competitive. Using traditional antitrust analysis measures, the major transpacific and transatlantic trade lanes remain unconcentrated and competitive. These trade lanes have a Herfindahi-Hirschman Index

[9] A Twenty-Foot Equivalent Unit (TEU) can be used to measure a ship's cargo carrying capacity. The dimensions of one TEU are equal to that of a standard 20 foot shipping container-20 feet long and 8 feet tall. Two TEUs are equal to one forty-foot-equivalent unit (FEU).

[10] Ocean common carriers that transport at least 0.1 percent market share or higher with a minimum of 18,000 containers per year in U.S. trades.

(HHI) of 835 and 1,354, respectively.[11] This also holds true when one further breaks out the transpacific trade by West Coast and East Coast, as well as the transatlantic Northern European trade. The other transatlantic trade lane, the Mediterranean, is moderately concentrated according to the index, although it is by far the smallest by volume of the noted trade lanes. None of the major trade lanes are highly concentrated using this measure.

Major Ocean Carriers by Operated Capacity

August 2011				**January 2018**			
Rank	**Operator**	**TEU**	**Share (%)**	**Rank**	**Operator**	**TEU**	**Share (%)**
1	Maersk	2,422,110	18.1	I	Maersk	4,151,807	22.7
2	MSC	2,016,197	15.1	2	MSC	3,147,525	17.2
3	CMA CGM	1,291,816	9.7	3	CMA CGM	2,514,170	13.7
4	COSCO	624353	4.7	4	COSCO	2,490,361	13.6
5	Hapag-Lloyd	622,799	4.7	5	Hapag-Lloyd	1,547,865	8.4
6	Evergreen	614,115	4.6	6	ONE	1,483,371	8.1
7	AM"	580,658	4.3	7	Evergreen	1,060,224	5.8
8	CSAV"	525,483	3.9	8	Yang Ming	594,806	3.2
9	I-fanjet"	511,661	3.8	9	PIL	384,807	2,1
10	CSCL"	51(1,958	3.8	10	Tim	366,292	2.0
II	MOL"	420,821	3.2	II	Hyundai (HMM)	347,136	1.9
12	OOCL	415,638	3.1	12	Wan Hai	235,591	1.3
13	NYK*	398,667	3.0		Total	18,323,955	100.0
14	Hamburg Sal"	394,652	3.0				
15	K Line"	342,763	2.6				
16	Zim	336,399	2.5				
17	Yang Ming	336,328	2.5				
18	Hyndai (HMM)	316,108	2.4				
19	PILL	265,919	2.0				
20	UASC*	234,815	1.8				
21	Wan Hai	171,423	1.3				
	Total	13,353,683	100.0				

* These carriers no longer exist as independent operating and marketing entities.

Data Source: ASX-Alphaliner Monthly Monitor, Top 30 Carriers, August 2011 & January 2018.

The global fleet has increased in size in recent years. At the beginning of 2018, ocean carriers deployed 21.1 million TEUs of ship capacity globally, a 70 percent increase from

[11] Concentration is assessed using the HHI. Theoretically, the greater the degree of market concentration and the fewer the competitors, the higher the HHI. In its merger guidelines, the Department of Justice's (DOJ) Antitrust Division regards markets as not concentrated if the HHI is below 1,500. Under DOJ guidelines, mergers, and other less problematical forms of horizontal collaborations, that do not result in concentrated markets are unlikely to produce adverse competitive effects and, ordinarily, do not require further government regulatory analysis.

2009. As the global fleet has increased, so has the capacity deployed on individual trade lanes. Overcapacity in nearly all trade lanes, including both major and minor lanes, has been an overarching theme for the container shipping industry in recent years, as carriers have increased capacity without a corresponding increase in demand. However, as consumer confidence and spending has grown, and the demand for ocean transportation services has increased, carriers have been able to fill their ships relatively close to capacity in the past year, despite having increased the total capacity on the major trade lanes.

Trade Lane	TEUs	HHI
Transpacific	22,594,058	835
Asia-WC	13,595,933	826
Asia-EC	8,907,327	943
Transatlantic	5,263,862	1,354
North Europe-U.S.	3,720,219	1,179
Mediterranean-U.S.	1,543,643	2,114

FMC monitoring data indicates that ocean carriers regularly experienced capacity utilization of over 90 percent on the inbound major transpacific trade throughout 2017 and about 90 percent on the transatlantic. Each of these trade lanes saw capacity utilization rise toward the end of 2017 compared to earlier in the year. However, vessel utilization on the backhaul route to Asia is only about 50 percent with only slightly higher levels returning to Europe. Although ships are sailing relatively full, rates have remained comparatively low and are 22 percent below their peak in 2010. When adjusted for inflation, real rates are down 31 percent since 2010. According to FMC monitoring data, rates have remained level on the major transatlantic trades.

There are some signs that the industry is moving towards a recovery from overcapacity and low freight rates. The percentage of the idled fleet has decreased. Many carriers have recently reported operating profits (i.e., covering operating costs but not necessarily covering capital costs). However, charter rates for vessels of all sizes remain substantially lower than their peaks prior to the recession. Additionally, there does not appear to be any indication that typical sailing speeds are increasing. Other factors that can affect moving to a recovery are continued economic growth in the United States and U.S. trade policies. However, an economic downturn and/or trade restrictions would have an adverse effect on demand for shipping and would slow down any recovery, thereby having a dampening effect on rates.

Nine of the remaining twelve ocean carriers are currently members of three global alliances-2M, OCEAN, and THE. These alliances -joint operating agreements of ocean carriers to discuss and agree on the supply of vessel capacity through the deployment of a specific service string or strings each operate services in the major transpacific (Asia-U.S. and Canada), transatlantic (Europe U.S. and Canada), and Asia-Europe trades and supply over 90 percent of the vessel capacity in each of these trade lanes. These three major alliances are not the only vessel sharing agreements in which these and other ocean carriers participate. Carriers can and do participate in multiple agreements filed at the FMC. These include various space charter agreements, vessel sharing agreements, vessel sharing alliances, joint service agreements, and cooperative working agreements. In addition to the three global vessel sharing alliances referenced above, ocean carriers participate in more than 325 agreements filed at the Commission.

Alliances can be very beneficial for U.S. exporters, importers, and consumers. Such alliances allow participants to obtain efficiencies and cost-savings that can be passed on to domestic consumers especially when healthy competition exists among vessel operators. Last, the benefits of alliances and other forms of joint commercial arrangements are recognized and addressed in the Shipping Act of 1984, as amended, and the contemporaneous Congressional record.[12]

A reassuring data trend show that even with the wave of mergers and acquisitions and new carrier alliance groupings, the individual ocean carriers within each alliance continue to independently and vigorously compete on pricing. Further, individual ocean carriers within the alliances continue to add and withdraw vessels from trades both inside and outside the alliances in which they participate, demonstrating that competition remains in both vessel capacity decisions and pricing decisions within the alliances. And over the last decade, the global vessel fleet has increased. The increase in capacity came from an increase in the number of vessels and an increase in the size of the new vessels entering the fleet. The increase in capacity occurred without a corresponding increase in cargo demand. Industry stakeholders have noted that the alternative to carrier alliances is further consolidation in the industry with fewer ocean carriers and less competition.

The Commission responded to the recent and ongoing structural changes in the international liner shipping industry with aggressive negotiations on proposed agreements and enhance monitoring programs. With the increased size and market share of carrier alliances over the last four years, the FMC has insisted on narrower authorities, more specific language, and enhanced monitoring requirements. For several years now with respect to these larger alliances, the FMC has required more clear and definite authority language for alliances. Monitoring for these large alliances, entailing more details and timely filing of monitor reports has increased.

As alliances are ongoing cooperative agreements rather than mergers, the Commission is charged by Congress with continuous monitoring after the initial review and following the effective date of the agreements. The Commission examines for

[12] The Conference Report for the Shipping Act of 1984, H. Rept. 600, states as follows: "Another important potential benefit to be considered is any efficiency-creating aspects of an agreement. Agreements involving significant carrier integration are, if properly limited to achieve such important benefits, to be favorably considered by the Commission and the courts. Joint ventures and other cooperative agreements can enable carriers to raise necessary capital, attain economies of scale, and rationalize their services. Pooling arrangements can also offer significant benefits in reducing excess capacity and promoting efficiency."

anticompetitive behaviors that would violate the Shipping Act. The Commission may challenge an agreement at any time after the effective date. Due to the constant monitoring role, industry and international trade expertise is important. The Commission is the expert agency on the ocean liner industry, dedicated to understanding the nuances of this important and complex industry.

Our expert analysts, economists, and attorneys maintain a careful watch on industry trends, being vigilant for any indications of anticompetitive behavior by the participants operating within the filed agreements. The Commission is diligent in monitoring economic conditions and carrier agreement activities to identify potential anti-competitive concerns and the possible need for Commission action. The FMC will continue to monitor industry trends to identify when the industry enters a full recovery and vessel supply/ cargo demand equilibrium. This will be important for analyzing the extent to which rate increases at that time are attributable to the recovery or to coordinated action by carriers.

The FMC prioritizes all filed agreements[13] on a *red-yellow-green* scale, with red being higher profile agreements with the highest probability of potentially adverse market effects based on the agreement's authority in combination with the underlying market. All global alliances are categorized as *red* agreements. For these alliances, FMC staff prepares scheduled briefings for management and conducts more detailed quarterly reviews. The FMC monitors these *red* agreements for any exercise of market power that could allow alliance members to raise and maintain prices above competitive levels.

The FMC conducts a four tiered-approach. The first tier is an immediate review of advance notifications of cancelled sailings or other changes in capacity that affect the supply of capacity of any individual alliance service by more than five percent of average weekly capacity. The second tier consists of a careful review of submitted minutes of most senior agreement committees that make vessel deployment decisions to assess the medium-to long-term outlook for capacity levels and how that could impact freight rates. Under the third tier, changes in individual alliance members' capacity and capacity projections and how that relates to changes in freight rates are analyzed. The final tier consists of reviewing and analyzing confidentially filed carrier data submitted by the alliances for completeness and accuracy to determine if this data reveals any potential red flags.

The Commission also monitors trends in other carrier and marine terminal operator agreement filings. It is important to note that carrier agreements containing rate discussion authority have experienced a steady decline in membership and a number have been terminated. More specifically, of the sixteen rate discussion agreements, five have been terminated entirely in the past few months, including the Transpacific Stabilization Agreement, or TSA, which has served as the primary price discussion forum for the

---

[13] At the end of FY 2017, there were 484 agreements on file.

ocean trade from Asia to the United States since 1989. Carriers appear to be ending their participation in rate discussion agreements for a number of reasons. Overcapacity continues to define the major east-west container shipping markets, keeping downward pressure on rates and limiting the effectiveness of these agreements. We also note carrier concerns over potential changes in the regulatory environment in the U.S. and abroad.

Further, the Commission monitors and analyzes commercial contracts confidentially filed in the FMC's SERVCON System between vessel-operating common carriers (VOCCs) and shippers for the transport of U.S. exports and imports. SERVCON is the Commission's repository for all filed service contracts, excluding exempt commodities, in the U.S. waterborne foreign commerce. Service contracts contain the rates, terms, and other service requirements agreed upon by VOCCs and shippers for the carriage of cargo. Commission staff conducts focused research and analysis on service contract terms and conditions, such as chassis usage/fees, demurrage terms/fees, etc., in order to investigate or clarify industry reports, gain better insight into emerging industry issues, and better inform policy decisions.

Review and analysis of confidentially filed commercial contracts between VOCCs and shippers provide a valuable tool to evaluate the competitive dynamics at play between shippers seeking to leverage cargo volumes in the pursuit of lower freight rates and/or special service terms and VOCCs competing to obtain that cargo.

FMC staff also systematically monitors a sampling of service contracts for a number of beneficial cargo owner and non-vessel-operating common carrier (NVOCC) shippers on an ongoing basis to track overall competitive conditions in various trades. These reviews are designed to protect the shipping public from unfair and deceptive carrier practices by identifying and addressing potential concerted carrier activity under filed agreements found to have resulted in discriminatory practices involving rates or charges applied to any locality, port, or persons due to those persons' status as shippers' association or ocean transportation intermediary.

Carriers are operating in an environment where trade volumes are growing at a more sluggish rate than has been historically typical of the sector. Concurrently, new vessels of the largest capacities continue to emerge from shipyards and be deployed into the world's trade lanes. All things being equal, and with no other shocks to the ocean carrier industry such as a spike in fuel costs or another shipping line bankruptcy, this is a formula for continued low rates, which ultimately benefits shippers, at least in the shorter term.

As noted earlier, although there has been a contraction in the number of lines operating in the international ocean trades, competition between companies remains vibrant and shippers continue to benefit from low rates. Overall market share of even the largest oceangoing carriers remain diffused. In the U.S. export and import trades combined, CMA CGM holds a 12.42 percent market share followed closely by Mediterranean Shipping Company (MSC) at 12.39 percent and Maersk in third position with 10.62 percent. These are far from "dominant" market positions as recognized by established economic standards. We will continue to look for any potential impact the carriers operating in the new alliances have on market dynamics, rates, and services.

While the United States' international trade depends on the liner trade, unfortunately there is no substantial U.S.-flag presence in the major transpacific and transatlantic trade lanes. The three largest carriers in the U.S. trades are CMA CGM, MSC, and Maersk Line. The invisible hand is not the only force that guides the global shipping industry, and nations throughout the world go to great lengths to support national companies, including saving them from bankruptcy. Some carriers receive government support, either directly or indirectly, or are owned in part by governments. The People's Republic of China (PRC) is the United States' largest trading partner in terms of cargo volume. The PRC actively invests in logistics, transportation, and infrastructure through initiatives such as Silk Road to advance strategic goals. The PRC-based COSCO Shipping and Hong Kong-based Orient Overseas Container Line (OOCL), will become the largest carrier of U.S. imports upon the two companies' planned 2018 merger. For the moment, such links between governments and national carriers can provide lower freight costs and greater service choices for imports and exports.

The ocean liner industry has been in a state of vessel oversupply for several years. The low freight rate structure in U.S. trade lanes is a direct reflection of that capacity supply/demand imbalance and American exporters and importers have been the beneficiary of those low freight rates. Such supply imbalances will not last forever. The Commission does not favor one competitor, sector, or industry stakeholder over another. We will continue to be attentive as we look for indications of rate increases that are products of market distorting, or collusive carrier business practices. However, it is important to remember that rate increases in and of themselves are not proof of an uncompetitive marketplace. At some point in the future, higher freight rates will be a normal result of a more equalized and healthy supply/demand marketplace.

The Commission continues to see marine terminal operators and port authorities' increased interest in how to use cooperative agreements filed with and reviewed by the Commission to their benefit. The nature and complexity of marine terminal operator agreements have increased considerably in recent years and marine terminal operators are cooperating in novel ways in an attempt to address the demands of significantly larger vessels unloading substantially larger numbers of containers at each port call. As a result, marine terminal operators have filed agreements to combine aspects of their operations, finance necessary infrastructure improvements, increase terminal velocity, develop collective solutions to mitigate cargo bottlenecks, and a host of other activities, all aimed at enhancing their ability to compete against other ports for cargo. There is a realization among these parties that seeking an alternate antitrust enforcement regime available to them through an agreement filed at the FMC can lead to increased efficiencies and lowered costs.

We would review with interest the application of any parties from the port and terminal sector who want to use agreements to achieve goals that ultimately benefit the American shipper and consumer. Due to the unique nature of these types of agreements, monitoring of terminal agreements is specifically tailored to the agreement's authority and potential competitive impact of the agreement.

Regulatory Reform and Agency Actions

Regulatory Reform

Throughout FY 2017 and into FY 2018, the Commission actively pursues ways to reduce unnecessary regulatory burdens and costs on the regulated community as mandated by the Shipping Act of 1984. Implementing Executive Order 13777, the Commission designated a Regulatory Reform Officer to lead a Regulatory Reform Task Force, charged with identifying outdated, cumbersome, or expensive rules that can be amended or eliminated. While the work of the Task Force is ongoing, the Commission has already taken steps to amend regulations related to Service Contracts, Negotiated Rate Agreements, and NVOCC Service Arrangements to eliminate or reduce unnecessary filing obligations. These changes will make it easier and more efficient for shippers and carriers to do business. Global supply chain operations will benefit through lower costs, which should result in savings realized by our U.S. exporters and importers.

On an ongoing basis, the Commission aggressively looks for ways to make compliance with Commission requirements easier and more cost effective for shippers, carriers, and ocean transportation intermediaries. We at the Commission are committed to identifying rules that are outdated or impede the efficient operation of business, and eliminating them whenever possible.

Relief from tariff publication requirements immediately comes to mind as another obligation ripe for Commission consideration. Currently, ocean transportation intermediaries and vessel- operating common carriers are required to publish both rates and rules in their tariffs, even though the overwhelming majority of cargo moving in most trade lanes does so under the terms of service contracts. In other words, current law and Commission regulations require companies to publish freight rates that have nothing to do with the actual market prices being charged to shippers. Continuing to mandate thousands of rate tariffs be published that do not reflect real market conditions, and have minimal, if any, use by industry participants when negotiating service contracts, is a requirement and expense that regulated entities could be relieved of under the exemption authority provided to the Commission by Congress.

Supply Chain Innovation Team Initiative

The Shipping Act contemplates a regulatory process for the foreign commerce of the United States with a minimum of regulatory costs. The Supply Chain Innovation Team Initiative (SCITI) was led by my colleague, Commissioner Rebecca F. Dye. The FMC initiative made a meaningful contribution towards enhancing supply chain efficiency for America's exporters and importers.

Whenever possible, the Commission seeks to facilitate the cooperation by stakeholders to develop non-regulatory commercial solutions to address bottlenecks in the international supply chain.

The SCITI was an outgrowth of the Commission's previous work on port congestion issues in the fall of2014. Launched in May of2016 and focused on challenges faced by America's international maritime supply chains, Commissioner Dye, with her volunteer teams of industry leaders composed of shippers, marine terminal operators, trucking companies, ocean carriers, port officials, labor representatives, logistics companies, and other stakeholders, worked to develop actionable commercial solutions - including in particular - the key content for a national seaport information portal that could provide the necessary critical information sought by all parties involved in moving containers to/from vessels, through seaports, and onward to a final destination.

SCITI created two teams one focused on import supply chains and the second focused on export supply chains. The work of both the import and export teams was summarized in a Final Report prepared by Commissioner Dye and presented to the Commission on December 5, 2017. *Supply Chain Innovation Teams Initiative: Final Report* presents the teams' view that greater visibility across the American freight delivery system was the one operational innovation likely to most increase U.S. international supply chain performance. The report also highlights the concept of a common National Seaport Information Portal for critical shipment information, possible organized by business dashboards tailored to the needs of each supply chain actor.

Petition P4-161Fact Finding 28

On December 7, 2016, the Coalition for Fair Port Practices, an organization of trade associations representing shippers, ocean transportation intermediaries, and domestic transportation companies, filed Petition (P4-16) asking the Commission to begin a new rulemaking proceeding to address practices by marine terminal operators (MTOs) and vessel operating common carriers (VOCCs) related to demurrage, detention, and related fees. Demurrage, detention, and related fees are charged by VOCCs and MTOs to compensate for the use of containers and terminal space and encourage the efficient movement of cargo through the terminals and the expeditious return of equipment. In P4-16, the petitioners claimed that the lack of any standards as to what constitutes unreasonable demurrage and detention practices under the Shipping Act of 1984 leads to unfair practices which undermine the integrity and efficiency of the U.S. ocean transportation system. The petitioners asked the Commission to issue a rule or alternatively a policy statement interpreting unreasonable demurrage and detention practices and provide industry with the tools it needs to more efficiently resolve demurrage and detention disputes.

The Commission received over a hundred comments on the Petition, and in January of this year held a two day public hearing that explored issues raised in the petition by soliciting testimony from shippers, ocean transportation intermediaries, ocean carriers, truckers, and marine terminal operators.

Based on the testimony heard in the public hearing, on March 5, 2018, the Commission voted to launch an investigation, headed by Commissioner Rebecca F. Dye, to examine practices of vessel operating common carriers and marine terminal operators related to detention, demurrage, and per diem charges. The investigation will focus on how demurrage and detention practices can optimize, not diminish, the performance of the American international freight delivery system. As the designated Investigative Officer, Commissioner Dye will have broad authority to conduct an investigation, including the power to issue subpoenas, to hold public and non-public sessions, and to require reports. Under this Order, she is charged with making recommendations for Commission action including investigations of prohibited acts; enforcement priorities; policies; rulemaking proceedings; or other actions warranted by the record developed in the proceeding. A final report of Commissioner Dye's findings and recommendations is due to the Commission no later than December 2, 2018.

Protecting the Public

In furthering our mission to protect the public from unlawful, unfair and deceptive practices, the Commission crossed an important milestone in FY 2017 with the successful launch of the Ocean Transportation Intermediary triennial update process. An important program in which the Federal Maritime Commission meets its mission of protecting the shipping public and American consumers from financial harm is by investigating and approving who the qualified and actual service providers are. In terms of non-vessel-operating common carriers and freight forwarders, what are called Ocean Transportation Intermediaries (OTIs), the Commission achieves that goal by licensing these entities.

Several years ago, the Commission reviewed a survey of OTIs and discovered that a substantial number had moved to new addresses without informing the FMC, that the person whose qualifications were reviewed as the basis of granting the license (Qualified Individual) was no longer an employee of the company, and several other filing discrepancies. A simple matter of not having the correct address of an OTI on file hampers the ability to have proper service in a legal matter and bringing our records up to date was an important goal.

Given advances in information technology, the Commission determined that there existed an opportunity to improve the quality and accuracy of information it has on file concerning OTIs while doing so in a manner that was easy to comply with, and had minimal burden. This is an online process where the responsible parties of an OTI update information related to ownership, corporate officers, business locations, changes in affiliation or branch office. Moving to a web- based update structure not only aids the Commission in meeting its mandate to safeguard the public, it significantly reduces the compliance burdens and costs upon the regulated entities. This completely online and no fee process takes approximately five minutes to complete.

FY 2019 Budget Request, Strategic Plan, Management Reforms

Fiscal Year 2019 Budget

The FMC is a small agency with a very technical mission and a need for a very specialized workforce. Our requested level of funding for FY 2019 is $27,490,000. Overall, the bulk of the Commission's budget, approximately 86 percent, is consumed by rent, salaries and benefits, and communications. Our staff includes a high percentage of transportation economists and attorneys career fields that tend to command more compensation in order to successfully recruit and retain qualified candidates and is the heart of the agency's mission. Overhead costs such as interagency services, commercial services, travel and transportation, supplies, and equipment account for most of the remaining budget dollars. The Commission has very little, if any, control over many of these costs. Year in and year out, the rent we are charged rises, the supplies and resources we purchase to support our economists and attorneys' competitive analysis and legal research cost more, and information technology (IT) costs-including IT security and telecommunications bills-rise. We constantly work to find a balance between our resources and our workload; however, if there is a surge of agreement filings, if a class of plaintiffs choose to seek relief at the FMC, or if our building security requirements increase, then we work to prioritize our mission-critical activity and reallocate resources to the extent possible.

Finding ways to conduct the Commission's business more efficiently is an important goal we share, Mr. Chairman. As such, the Commission works to find ways to make every dollar appropriated to us go as far as it can. A recent example of innovative cost-sharing is our agreement with another small, independent agency, the Surface Transportation Board (STB), to share the services and cost of a single Equal Employment Opportunity (EEO) Officer to ensure both agencies' responsibilities while maintaining solid support of our EEO principles.

As I mentioned earlier in this testimony, the Federal Maritime Commission continues to faithfully implement the purposes and mission of the Shipping Act. I am proud of the work the Commission's staff does each day to ensure a competitive and reliable international ocean transportation supply system that supports the U.S. economy and protects the public from unfair and deceptive practices.

Strategic Plan for 2018-2022

A proven method of achieving strong performance at an organizational level is through focused and meaningful strategic planning. Strategic planning is a driving force in an organization's success. Government agencies benefit from strategic planning that is focused, and designed to unite all agency team members to find ways to achieve our mission more effectively while delivering value to the taxpayer.

Early this year, the Commission finalized a new Strategic Plan for FY 2018-2022. This document will guide our work into the future as the Commission continues its important work that supports the U.S. economy and protects the public from unfair and deceptive practices.

Agency Reform and Long-Term Workforce Plan

The President has made reshaping the Federal government one of the key initiatives of his Administration. Through an Executive Order issued in March and a Memorandum issued in April 2017 by the Director of the Office of Management and Budget (OMB), the Administration instructed departments and agencies throughout the Federal government to include an Agency Reform and Long-Term Workforce Plan (Workforce Plan) as part of their FY 2019 budget submissions. A prime directive in the Executive Order and OMB Memorandum was for Federal agencies to explore, develop and implement plans to streamline, consolidate and flatten their organizational operations and structure.

Over the last year, the Commission developed a Workforce Plan as directed by OMB. In broad terms, our 5-year Workforce Plan will: reduce the number of SES positions; reduce the number of supervisory positions; establish a new two-tier SES structure to realign and control SES salary costs; realign and combine functions within the Commission (subject to Congressional approval); and continue our emphasis on achieving operational efficiencies and improving customer service through automation projects.

Our goal is to find ways to do more while controlling costs. Delayered work groups with broader spans of control and less hierarchy have been proven to improve both efficiency and employee engagement. We are working to reshape the FMC and improve operational effectiveness as required by the Administration while minimizing the impact to the 120 committed and vital employees of the FMC.

Conclusion

Thank you for this opportunity to discuss the mission of the Federal Maritime Commission, current state and future challenges of the ocean shipping industry, as well as highlight some of the Commission's recent achievements and future priorities. I am always ready to be of any assistance to the Subcommittee and its Members. Thank you for your attention and I will be pleased to answer any questions you may have.

INDEX

A

B

C

D

E

F

G

H

R

S

T

Coast Guard: Maritime Safety, Transparency and Acquisitions

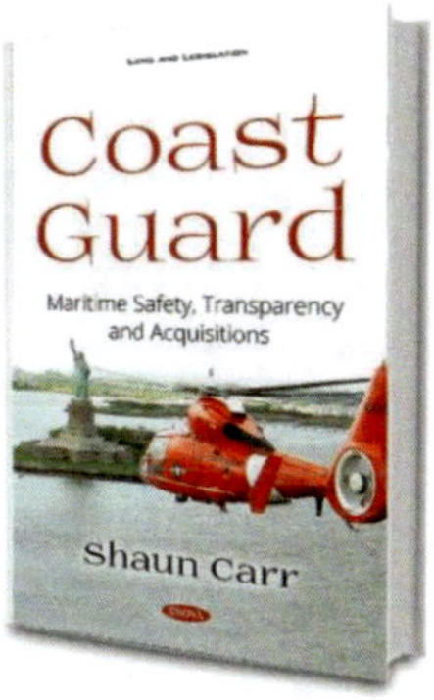

Editor: Shaun Carr

Series: Laws and Legislation

Book Description: The Coast Guard's responsibilities are specified in legislation establishing the agency as well as authorization bills typically passed by Congress every one to two years and in Department of Homeland Security appropriations acts.

Hardcover ISBN: 978-1-53614-891-6
Retail Price: $230

Female Officers in the Department of Defense and Coast Guard: Increasing Recruitment and Accessions

Editors: Patricia Vaughn

Series: Defense, Security and Strategies

Book Description: This book evaluates the extent to which accessions of women into the officer corps have increased, and DOD and the Coast Guard have determined resources and funding to increase the accessions of women into the officer corps; and DOD and the Coast Guard have initiatives and an oversight framework to increase the recruitment and accessions of female officers.

Softcover ISBN: 978-1-63485-397-2
Retail Price: $95